USING THE BUILDING REGULATIONS

Part M Access

USING THE BUILDING REGULATIONS

Part M Access

Melanie Smith

AMSTERDAM • BOSTON • HEIDELBERG • LONDON • OXFORD • NEW YORK
• PARIS • SAN DIEGO • SAN FRANCISCO • SINGAPORE • SYDNEY • TOKYO
Butterworth-Heinemann is an imprint of Elsevier

Butterworth-Heinemann
An imprint of Elsevier
30 Corporate Drive, Burlington, MA 01803

First published 2006

British Library Cataloguing in Publication Data
A catalogue record for this book is available from the British Library

Library of Congress Cataloguing in Publication Data
A catalogue record for this book is available from the Library of Congress

ISBN 0 7506 6450 9

For information on all Elsevier Butterworth-Heinemann publications visit our website at www.books.elsevier.com

Typeset by Macmillan India Ltd, Bangalore, India

Printed and bound in United Kingdom

Author Biography

Melanie Smith is a senior lecturer in the School of the Built Environment at Leeds Metropolitan University. She is a Chartered Building Surveyor and Fellow of the Royal Institution of Chartered Surveyors. She sits on the RICS DDA Working Party and is a consultant member of the National Register of Access Consultants. Having gained a first class honours degree in Building Technology from UMIST, she worked as a Building Control Officer with Leeds City Council, and set up her consultancy business in 1988.

M J Billington (Series Editor) is a Chartered Building Surveyor. He has a lifetime's involvement in the construction industry having worked at one time or another in design, construction and control (both private and public sectors). He was formerly Senior Lecturer in building control and construction at De Montfort University, Leicester before leaving to join the private sector, where he continued to act as visiting lecturer at a number of universities. He has published many technical papers and a number of books on building regulations and building defects and is a contributor to Knight's Guide to Building Control Law and Practice. Currently, he is Managing Director of Construction Auditing Services Ltd, a company that specialises in latent defects insurance technical auditing.

Contents

List of Figures

List of Tables

Preface

Part M of the Building Regulations is concerned with the access to and use of buildings. Within the context of the Building Regulations, it aims for buildings, newly built or newly altered, to provide reasonable access to get to and into the building, and to be able to use its facilities.

This book provides a good informative tool for the designer to be able to meet the requirements. It explains the provisions included in the Approved Document M to the Regulations. It gives background information, and explains the relationship of Part M with the British Standard BS8300, the Disability Discrimination Act (DDA) and what an Access Statement is.

Melanie Smith
MJ Billington

Acknowledgements

We wish to thank Ben Cottle, building surveyor, and to acknowledge his considerable help in the preparation of this book.

We also thank Dinah Everard and Nathan Legge for their input and advice, and members of the RICS DDA Working Party.

Acknowledgements and thanks must go to:

Nick Haddon and The Random House Group Ltd for permission to reproduce extracts from 'The Curious Incident of the Dog in the Night Time';

The Navigator Group Ltd for permission to use their Ten Point Plan; and

Her Majesty's Stationery Office, Norwich, for permission to reproduce the robust construction detail for level thresholds.

And many thanks to Kathleen Smith, Rowena Hanson-Abbott and Rachel and Sari Legge for expanding our knowledge of challenges people face.

About this series of books

Whether we like it or not, the Building Regulations and their associated Government-approved guidance documents get more complex with every update, often requiring the services of specialist professionals (services engineers, fire engineers etc.) to make sense of the provisions. New areas of control are being introduced each year and the scope of the existing regulations is being extended with each revision.

The technical guidance given in the current Approved Documents is only of use in the design of extremely simple and straightforward buildings using mainly traditional techniques. For larger and more complex buildings it is usually better (and more efficient in terms of building design) to use other sources of guidance (British and European Standards, Building Research Establishment Reports etc.), and although a great many of these other source documents are referenced in the Approved Documents no details of their contents or advantages of use are given.

The current Approved Documents usually fail to provide sufficient guidance just when it is needed, i.e. when it is proposed to deviate from the simple solutions or attempt to design something slightly unusual, thus encouraging adherence to traditional and (perhaps) unimaginative designs and details, and discouraging innovation in the majority of building designs.

This series of books, by addressing different parts of the Building Regulations in separate volumes, will enable each Part to be explored in detail.

The information contained in the Approved Documents is expanded not only by describing the traditional approach but also by making extensive reference to other sources of guidance contained in them. These 'alternative approaches' (as they are called in the Approved Documents) are analysed and the most critical parts of them are presented in the text with indications of where they can be used to advantage (over the traditional approach).

As this is a new concept in building control publications our aim is to develop the series by including examples of radical design solutions that go beyond the Approved Document guidance but still comply with the Regulations. Such innovative buildings already exist, one example being the Queen's Building at De Montfort University in Leicester, which makes extensive use of passive stack ventilation instead of traditional opening windows or air conditioning.

About this book

This book presents a detailed analysis of Part M – Access and use – of Schedule 1 to the Building Regulations 2000. Before publication of the latest revision of Part M and its accompanying Approved Document in 2004, the final section of the Disability Discrimination Act 1995 became law and a new British Standard, BS 8300:2001 *Design of buildings and their approaches to meet the needs of disabled people*, was published. This comprehensive code of practice became the main reference document when considering access for disabled people. The information contained in it went far beyond that in the 1999 edition of Approved Document M. However, the Standard does not limit itself to only those matters properly addressed by Building Regulations and covers many aspects of design that are not the subject of Regulation and therefore not included in the Approved Document.

After the publication of BS8300:2001, the guidance in Approved Document M seemed somewhat lacking and out of date. The 2004 edition of Approved Document M has been altered and extended from the 1999 edition so that is again a first reference source. It has incorporated much of the guidance in BS8300, and in some instances gone beyond the recommendations in that document.

Another important change is the introduction of Access Statements. The Approved Document recommends that Access Statements accompany applications to 'identify the philosophy and approach to inclusive design adopted'. This is particularly important, and a sensible concept when the proposals differ from the recommendations given in the Approved Document.

As a result of the changes in the law much confusion exists over the relative status of the guidance contained in the Approved Document and BS 8300, and the requirements of the Disability Discrimination Act. This book discusses and explains the contents of Approved Document M to meet the requirements of Part M, shows the relationship between Part M of Schedule 1 to the Building Regulations and the Disability Discrimination Act 1995, and demonstrates how appropriate use of other guidance, such as BS8300, can be to the advantage of the designer or developer.

This book is aimed at designers, builders, students on construction and building surveying related courses, building control professionals and anyone else with an interest in the built environment. Its purpose is to keep them better informed and more able to deal with a complex and evolving area of law which directly affects everyone.

1

Series introduction

1.1 Introduction

Although we may not be aware of it, the influence of the Building Regulations is around us all of the time.

In our homes, building regulations affect and control the:

- size and method of construction of foundations, walls (both internal and external), floors, roofs and chimneys
- size and position of stairs, room exits, corridors and external doors
- number, position, size and form of construction of windows and external doors (including glazing)
- methods for disposing of solid waste
- design, construction and use of the services such as:
  - above and below ground foul drainage taking the waste from kitchen and bathroom appliances (including the design and siting of the appliances themselves)
  - rainwater disposal systems including gutters and downpipes from roofs and drainage from paths and paving
  - electrical installations
  - heating and hot water installations using gas, oil or solid fuel
  - fire detection and alarm systems
  - mechanical ventilation systems
- design and construction of the paths outside the house that:
  - lead to the main entrance, and
  - are used to access the place where refuse is stored.

In a similar manner, they also affect the places where people go when away from their homes such as:

- factories, offices, warehouses, shops and multi-storey car parks
- schools, universities and colleges

- leisure, sport and recreation centres
- hospitals, clinics, doctors surgeries, health care centres and other health care premises
- hotels, motels, guest houses, boarding houses, hostels and halls of residence
- theatres, cinemas, concert halls and other entertainment buildings
- churches and other places or worship.

In fact, anything that can normally be considered to be a building will be affected by building regulations. But it is not just the design and construction of the building itself that is controlled.

The regulations also affect the site on which the building is placed in order to:

- lessen the effect of fire spread between neighbouring buildings
- permit access across the site for the fire brigade in the event of fire
- allow access for disabled people who may need to get from a parking place or site entrance to the building, and
- permit access for refuse collection.

1.2 What are the Building Regulations?

When asked this question most people (assuming that they have even heard of the regulations) will usually bring to mind a series of A4 documents with green and white covers and the words "Approved Document" on the front! These documents are not, of course, the Building Regulations, but have come to be regarded as such by most builders, designers and their clients, and it is this misconception that has led to a great deal of confusion regarding the true nature of the building control system and the regulations. When applied to England and Wales, the Building Regulations consist of a set of rules that can only be made by Parliament for a number of specific purposes. The purposes include:

- ensuring the health, safety, welfare and convenience of persons in or about buildings and of others who may be affected by buildings or matters connected with buildings,
- furthering the conservation of fuel and power, and
- preventing waste, undue consumption, misuse or contamination of water.

The regulations may be made 'with respect to the design and construction of buildings and the provision of services, fittings and equipment in or in connection with buildings'.

Originally (in Victorian times), the regulations (or byelaws as they were known then) were concerned only with public health and safety, but in the late twentieth century additional reasons for making building regulations were added so that it would now seem possible to include almost anything under the banner of 'welfare and convenience'.

The Regulations are of two types:

(a) those that deal with issues of procedure or administration such as:

- the types of work to which the regulations apply
- the method of making an application to ensure compliance and the information that must be supplied to the controlling authority
- the frequencies and stages at which the control authorities must be informed of the work
- details of the testing and sampling that may be carried out by the controlling authorities to confirm compliance
- what sorts of work might be exempted from regulation control
- what can be done in the event of the work not complying with the regulations

(b) those that describe the 'standards' which must be met by the building (called 'substantive' requirements) such as:

- the ability of the building to:
  - retain its structural integrity
  - resist the effects of fire and allow people to escape if a fire should occur
  - resist dampness and the effects of condensation
  - resist the passage of sound
  - minimize the production of carbon dioxide by being energy efficient
  - be safe to use, especially where hazards of design or construction might exist, such as on stairways and landings or in the use of glass in windows, doors or as guarding
  - maintain a healthy internal environment by means of adequate ventilation.
- the safe installation and use of the building's services including:
  - electric power and lighting
  - boilers, open fires, chimneys, hearths and flues
  - unvented heating and hot water systems
  - sanitary installations and above and below ground drainage
  - foul and waste disposal systems
  - mechanical ventilation and air conditioning systems
  - lifts and conveyors.

Because the regulations are phrased in functional terms (i.e. they state what must be achieved without saying how this must be done), they contain no practical guidance regarding methods of compliance. The intention of this approach is that it gives designers and builders flexibility in the way they comply, and it does not prevent the development and use of innovative solutions and new materials and methods of construction. Of course, much building work is done in traditional materials using standard solutions developed over many years and based on sound building practice. To assist designers and contractors in these accepted methods, the Government has provided non-mandatory guidance principally in the form of 'Approved Documents', there being an Approved Document that deals with each substantive provision of the Building Regulations. This does not prevent the use of other 'official' documents such as Harmonised Standards (British or European), and the adoption of other methods of demonstrating compliance such as past experience of successful use, test evidence, calculations, compliance with European Technical Approvals, the use of CE-marked materials etc.

1.3 How are the Regulations administered?

For most types of building work (new build, extensions, alterations and some use changes), builders and developers are required by law to ensure that they comply with the Regulations. At present this must be demonstrated by means of an independent check that compliance has been sought and achieved.

For this purpose, building control is provided by two competing bodies – Local Authorities and Approved Inspectors.

Both Building Control Bodies will charge for their services. They may offer advice before work is started, and both will check plans of the proposed work and carry out site inspections during the construction process to ensure compliance with the statutory requirements of the Building Regulations.

1.3.1 Local authority building control

Each Local Authority in England and Wales (Unitary, District and London Boroughs in England and County and County Borough Councils in Wales) has a Building Control section. The Local Authority has a general duty to see that building work complies with the Building Regulations unless it is formally under the control of an Approved Inspector.

Individual local authorities co-ordinate their services regionally and nationally (and provide a range of national approval schemes) via LABC Services. Full details of each local authority (contact details, geographical area covered, etc.) can be found at www.labc-services.co.uk.

1.3.2 Approved inspectors

Approved Inspectors are companies or individuals authorized under sections 47 to 58 of the Building Act 1984 to carry out building control work in England and Wales.

The Construction Industry Council (CIC) is responsible for deciding all applications for approved inspector status. A list of approved inspectors can be viewed at the Association of Consultant Approved Inspectors (ACAI) web site at www.acai.org.uk.

Full details of the administrative provisions for both local authorities and approved inspectors may be found in Chapter 5.

1.4 Why are the Building Regulations needed?

1.4.1 Control of public health and safety

The current system of building control by means of Government regulation has its roots in the mid-Victorian era. It was originally set up to counteract the truly horrific living and working conditions of the poor working classes who had flocked to the new industrial towns in the forlorn hope of making a better living. Chapter 2 of the first book in this series (*Using the Building Regulations – Administrative Procedures,* published by Elsevier Butterworth Heinemann, ISBN 0 7506 6257 3) describes the factors which caused this exodus from the countryside and the conditions experienced by the incomers; factors which led to overcrowding, desperately insanitary living conditions and the rapid outbreak and spread of disease and infection. There is no doubt that a punitive system of control was needed at that time for the control of new housing, and the enforcement powers given to local authorities (coupled with legislation that dealt with existing sub-standard housing) enabled the worst conditions to be eradicated and the spread of disease to be substantially halted.

The Victorian system of control based purely on issues of public health and safety enforced by local authorities continued to be effective for the next 100 years, the only major change being the conversion of the system from local byelaws to national regulations in 1966.

1.4.2 Welfare and convenience and other controls

The first hint of an extension of the system from one based solely on public health and safety came with the passing of the Health and Safety at Work, etc. Act in 1974 (the 1974 Act). Part III of the 1974 Act was devoted entirely to changes in the building control system and regulations and it increased the range of powers given to the Secretary of State. Section 61 of the 1974 Act enabled him to make regulations for the purposes of securing the welfare and convenience (in addition to health and safety) of persons in or about buildings. Regulations could also be made now for furthering the conservation of fuel and power and for preventing the waste, undue consumption, misuse or contamination of water. The 1974 Act was later repealed and its main parts were subsumed into the Building Act 1984.

1.4.3 The new system and the extension of control

Initially, the new powers remained largely unused and it was not until the coming into operation of the completely revamped building control system brought about by the 1984 Act and the Building Regulations 1985 that the old health and safety-based approach began to change. The 1984 Act also permitted the building control system to be administered by private individuals and corporate (i.e. non-local authority) bodies called Approved Inspectors in competition with local authorities, although enforcement powers remained with local authorities. The new powers have resulted in the following major extensions of control to:

- heating, hot and cold water, mechanical ventilation and air conditioning systems
- airtightness of buildings
- prevention of leakage of oil storage systems
- protection of LPG storage systems
- drainage of paths and paving
- access and facilities for disabled people in buildings (although the reference to disabled people has now been dropped)
- provision of information on the operation and maintenance of services controlled under the regulations
- measures to alleviate the effects of flooding in buildings
- measures to reduce the transmission of sound within dwellings and between rooms used for residential purposes in buildings other than dwellings.

Furthermore, a recent consultation in 2004 put forward proposals intended to facilitate the distribution of electronic communication services (Broadband) around buildings in a proposed Part Q, presumably under the banner of convenience.

As the scope of control has increased, the Government has attempted to simplify the bureaucratic processes that this increase would undoubtedly lead to by allowing much work of a minor nature and/or to service installations to be certified as complying by a suitably qualified person (e.g. one who belongs to a particular trade body, professional institution or other approved body).

1.4.4 The future of building control in England and Wales

This section derives its title from a Government White Paper (Cmnd 8179) published in February 1981. In the second paragraph of this document, the Secretary of State set out the criteria which any new building control arrangements would be required to satisfy. These were:

- maximum self-regulation
- minimum Government interference
- total self-financing
- simplicity in operation.

One out of four (total self-financing) may not seem to be a particularly good result, and it has often been the case that the average local authority building control officer has been inadequately prepared through inappropriate education and training to take on the task of assessing compliance with many of the regulation changes listed above. It has been claimed that this problem has been solved by the introduction of Approved Inspectors onto the building control scene. Since these are staffed almost entirely by ex-local authority building control officers, it would seem that the net result of the partial privatization of building control has only been to redistribute a finite number of similar people without any improvement in education or training, although the adoption of a more commercial attitude by Approved Inspectors may be a good or bad thing depending on your point of view.

It seems almost inevitable (without a change of Government or in Government thinking) that the areas of control will increase and that more 'suitably qualified people' will be entitled to certify work as complying with the regulations. It is also likely that local authorities will remain as the final arbiter in matters of enforcement although it is likely that their direct involvement in day-to-day building control matters will diminish, to be taken over by the private sector. Indeed, most building control work on new housing is already dealt with by the private sector.

Although the broad subject area covered by the Building Regulations is roughly the same across the European Union (and in former British colonies such as Canada, Australia and New Zealand), the main difference between the system in England and Wales and that in other countries lies in the administrative processes designed to ensure compliance. Our mix of control mechanisms encompassing both public (local authority) and private (approved inspector) building control bodies offers choice but also potential conflict. The system is further complicated by the existence of certain 'self-certification' schemes for the installation of, for example, replacement windows and doors or combustion appliances, and some work, which has to comply with the regulations, but is 'non-notifiable' if carried out by a suitably qualified person.

In fact, we are the only country in the EU with such a 'mixed economy'. Most countries (Scotland and Northern Ireland, Denmark, The Netherlands, the Irish Republic, etc.) use a system run exclusively by the local authority. In Sweden, the building control system was privatized in 1995 so that the work of plan checking and site inspections is carried out by a suitably qualified 'quality control supervisor' employed by the building owner, although the local authority still has to be satisfied that the work is being properly supervised and may carry out spot checks and inspections to confirm this.

Some years ago, the UK government consulted on proposals to extend control of work governed by the Building Regulations to a range of bodies (some of which could be engaged in design and construction), provided that they were suitably qualified and insured. This would mean, for example, that a firm of architects would be able to take complete control of their own building control processes for work that they had designed without using a local authority or Approved Inspector. Such a market-led system would seem to be in accordance with all the aims listed at the beginning of this section and provided that the necessary safeguards could be put in place to prevent corruption and build public confidence, it would seem to be a sensible way forward. The consultation exercise did not result in any companies being approved to control their own work although the current system of certification of compliance by suitably qualified persons did come out of the exercise. Whether this was caused by political interference, objections from the building control establishment or lack of confidence by companies who still wanted the comfort of a third party to do their regulation checking for them, is not known.

2

General considerations

2.1 Introduction

Each of us falls short of that non-existent 'ideal' concept of modular man to some extent, be it in girth, height, stamina, understanding, ability, hearing or sight. We are all 'disabled' and it is no longer acceptable to marginalize people who happen to be further from that concept than others. The Office of National Statistics (2002) reports that 8% of all adults have difficulties using doors. The Disability Rights Commission estimates that 1 in 5 adults has a disability. The demographics of ageing predict that by 2030, the population over 65 will have doubled and those over 80 will have trebled (Hanson, 2004). Problems caused by inaccessible environments affects or are likely to affect all of us. We need to address this.

As explained in Chapter 1, the Building Regulations (when applied to England and Wales) consist of a set of statutory requirements that can only be made by Parliament and which must be met when carrying out building work. Building work is defined in the Regulations and can consist of:

- the erection or extension of a building;
- the provision or extension of a controlled service or fitting in or in connection with a building;
- the material alteration of a building, or a controlled service or fitting;
- work relating to a material change of use;
- the insertion of insulating material into the cavity wall of a building; and
- work involving the underpinning of a building.

The substantive requirements of the Building Regulations are contained in Schedule 1 and are divided into different Parts, currently Parts A – P (with no Part I or Part 0).

To assist designers and developers, the Office of the Deputy Prime Minister publishes Approved Documents, which set out recommendations showing various ways of complying with the mandatory regulations. It should be stressed that the recommendations given in the Approved Documents are just that. There is no legal obligation to use the Approved Documents when carrying out a building project to which the regulations apply. The applicant is perfectly entitled to choose whether or not to use all or some of the relevant Approved Documents or indeed, some parts of them. However, should an alternative solution be adopted, the onus will be on the designer/developer to establish that the requirements have been met in some other way and he or she may be called upon by the building control body to demonstrate this.

There is at least one Approved Document for each Part although Part L, Conservation of fuel and power, is split into Part L1 – Dwellings, and Part L2 – Buildings other than dwellings, and has an Approved Document L1 and an Approved Document L2. Part M of Schedule 1 to the Building Regulations covers access to and use of buildings, and is commonly known as the Part associated with access for disabled people.

The Regulations are revised from time to time and the Approved Documents are amended and updated, including complete revisions at regular intervals. The latest revision of Part M and its Approved Document is the 2004 edition. As explained above, the guidance in the Approved Document, although giving the best practice advice at the time of its publication, is not the only guidance available for complying with regulations. Other acceptable guidance could be incorporated in British, European or International Standards, academic research publications, and published materials from experts in the field, etc. When the guidance in Approved Document M is not used, the designer could produce an Access Statement to show the concepts and guidance used. Guidance on the production of these is included in this book.

Before the latest revision of Approved Document M was published, the final section of the Disability Discrimination Act 1995 became law and a new British Standard, BS 8300:2001 *Design of buildings and their approaches to meet the needs of disabled people*, was published. This comprehensive code of practice became the main reference document when considering access for disabled people. The information contained in it went far beyond that in the 1999 edition of Approved Document M. However, the Standard does not limit itself to only those matters properly addressed by Building Regulations, it covers many aspects of design that are not the subject of Regulation and therefore not included in the Approved Document.

After the publication of BS8300:2001, the guidance in Approved Document M seemed somewhat lacking and out of date. The 2004 edition of Approved Document M has been altered and extended from the 1999 edition, so that is again a first reference source. It has incorporated much of the guidance in BS8300, and in some instances gone beyond the recommendations in that document.

One of the most impressive changes is the title of the 2004 edition of Part M and its Approved Document M which is now entitled '*Access to and use of buildings*'. There is no mention of 'disabled people'. The aim is to create an inclusive approach so that the design of new buildings and building work accommodates the needs of all people. It is not only 'disabled' people who find some aspects of the built environment challenging and the Building Regulations now reflect this.

The second most important change is the introduction of Access Statements. The Approved Document recommends that Access Statements accompany applications to "identify the philosophy and approach to inclusive design adopted". This is particularly important, and a sensible concept when the proposals differ from the recommendations given in the Approved Document.

This book will discuss and explain the content of Approved Document M to meet the requirements of Part M, show the relationship between Part M of Schedule 1 to the Building Regulations and the Disability Discrimination Act 1995, and appropriate use of other guidance.

Approved Document M is often referred to as ADM throughout this book and the Disability Discrimination Act 1995 as the DDA. Of necessity much of the content will bear a remarkable similarity to the Approved Document without the courtesy of an acknowledgement as to the source. The authors therefore acknowledge here Approved Document M as the source of much material in this book, direct the reader to consult the Document, and commend this book for further explanations, background knowledge, understanding and recommendations.

2.2 What are the requirements?

Building Regulations are not applied retrospectively and compliance is only needed when new or altered building work is taking place or when there is a change of use. Not every regulation applies in every case. The designer should consult the actual Regulations, particularly Regulations 3–6 (see Chapter 4 of the first book in this series – *Using the Building Regulations – Administrative Procedures,* published by Elsevier Butterworth Heinemann ISBN 0 7506 6257 3), to determine exactly what applies in a particular circumstance. Further advice is given later in this chapter. However, at this point it would be useful to consider what provisions are required by Part M of Schedule 1 to the Building Regulations. Where Part M applies, the Regulations require that reasonable provision should be made for the following:

Buildings other than dwellings

(a) so that people can reach the principal entrance to the building, and other entrances as required, from the site boundary, from car parking within the site, and from other buildings on the same site regardless of ability, disability, age or gender;

(b) so that elements of a building do not constitute a hazard to users, especially those with impaired sight, but rather assist in way-finding;
(c) so that people, regardless of ability, age or gender can have access into, and within, any storey of the building and to the building's facilities, subject to the usual gender-related conventions of sanitary conveniences;
(d) for suitable accommodation in audience or spectator seating for people in wheelchairs, or for people with other disabilities;
(e) for aids to communication in auditoria, meeting rooms, reception areas, ticket offices and information points for people with sight or hearing impairments;
(f) for sanitary accommodations for the users of the building.

Dwellings

(a) so that people can reach the principal entrance of the dwelling, or a suitable alternative, from the point of access (see 'Interpretation – Dwellings only' below);
(b) so that people can gain access into the dwelling and move easily around its principal storey, unimpeded;
(c) for a WC provision in the entrance storey where this contains habitable rooms, or no higher than the principal storey.

2.3 Chapter layouts

This book therefore covers guidance and recommendations pertinent to the above requirements and generally to access provisions. To design well, the designer needs to consider the reasons behind the requirements and the importance of measurements given. Blind adherence to the provisions in the approved document may be acceptable, but too many completed designs have shown that minimum space requirements have not been met. This can result in, for example, an expensive WC cubicle supposedly for use by wheelchair users being unusable by them. The remainder of this chapter therefore explains some of the challenges people face when negotiating their way around the built environment, and gives some practical measurements of the space that people need in order to help the designer.

An explanation of terms is useful, and interpretation of common words and phrases encountered is included.

Having established, what aspects of design we are controlling and what the needs of real people relating to these are, we can look again at the regulations and see what is being controlled and in what circumstances. Chapter 2 concludes with a discussion of historic properties, because it is often thought that by being historic such buildings do not need to comply with the provisions of the Building Regulations or the DDA. This is not the case, but the different challenges of special historic features and the specific needs of people to gain access and use the building need to be reconciled if there is any conflict, in sensitive and sensible ways. The use of access statements is most helpful here.

Access statements are relatively new and therefore Chapter 3 discusses these, giving some of the advice offered by the Disability Rights Commission as well as the recommendations given in the Approved Document. Each Statement will be individual and pertinent to the work in question for which Building Regulation compliance is sought.

Access to and in buildings is covered in detail in this book in two main chapters: Chapter 4 Buildings other than dwellings; and Chapter 5 Dwellings. The sub-headings in these chapters follow the route a person would take when travelling to the building, entering it, manoeuvring around the building and using any of its facilities. The requirements of the Building Regulations and the recommendations given in the Approved Document are given, as well as any further guidance offered by such publications as BS 8300 (BSI, 2001).

Chapter 6 discusses the requirements of the Disability Discrimination Act 1995 and the use of other publications such as BS8300, the British Standard Code of Practice, *Design of buildings and their approaches to meet the needs of disabled people* (BSI, 2001).

Chapter 6 goes on to explain the Disability Discrimination Act 1995. The legal framework of this legislation is explained along with its relationship with the Building Regulations and Part M. Although these are separate pieces of legislation, there are practical implications of working with both. There is a comparison between Building Regulation Part M and the DDA, and their relationship is discussed.

As people expect and are expected to be able to gain access to buildings, so should they also be afforded egress. The most important provision for this is means of escape in case of fire or other emergency, and to be alerted to the presence of the fire. The current guidance for this in new buildings and for new work is contained in Approved Document B with further reference in the BS 5588 series (BSI, 1999). Provision for people of impaired mobility and other disabilities is especially contained in BS 5588-8:1999, '*Fire precautions in the design and use of buildings, Code of practice for means of escape for disabled people*'. This is included in the discussions in Chapter 6.

A major tool for avoiding discrimination in the design of a building and its environs is an Access Audit. These are discussed in Chapter 6 and an example shown in Appendix 2. An access audit used for the DDA is different to an access statement as required by Building Regulation and Planning Approval.

2.4 People and poor design and detailing

Part M of Schedule 1 to the Building Regulations has been produced so that reasonable provision is made to ensure that buildings are accessible and usable by people. The implication being that they are accessible and usable by all people.

People come in many shapes and sizes. When considering reasonable access to buildings and their facilities, these different shapes and sizes must be considered. An architectural-award winning design is not viable if it does not accommodate ordinary people. Anthropometrical and ergonomical research is improving our knowledge of the extents and limits of different people, and the results of much of this research are now used in the codes and guidance available, including ADM. There are no perfect single solutions for access, as people's abilities or requirements can be at opposite extremes. For example, a very tall person would need a door handle within his or her reach, while a very small person would probably find what is comfortable for this person is too high for them. A wheelchair user would like a ramp to enable them to get up an incline, while a person with severe arthritis who has great difficulty bending at the ankle may find a ramp too painful and would prefer well-designed steps. Different measures to suit different people therefore may have to run alongside each other.

It may not be possible or reasonable to consider all the different types and ranges of ability, so the designer often considers common features of ability and designs for these. There is no definition of 'disabled person' or 'disability' in Part M or ADM, as there is no need. This Part of the Building Regulations is not concerned with access for disabled people but with access for people. It defines 'accessible' as '*Accessible, with regard to buildings or parts of buildings, means that people, regardless of disability, age or gender, are able to gain access*'. As always with the regulations this is tempered with reasonableness, and it also does not state that everyone should be able to, or be entitled to, gain access everywhere. However a rule of thumb could be that where other people can physically go, then anybody should be physically able to go.

The main features of people's abilities are often divided up as follows, although any one person could have more than one of the features, and could have them to a lesser or greater extent:

- **Mobility impairments** – for example requiring the use of walking aids such as crutches, sticks, walking frames, electric wheelchairs, ordinary wheelchairs, walking trolleys, etc. The term *ambulant disabled* can be used to describe people who have mobility problems, but the term ambulant disabled more accurately describes people who may have disabilities but can walk and this includes people who may have hearing, sight or dexterity impairments as well as those with mobility problems.

- **Dexterity impairments** – for example, which result in an inability to grip small items, or to easily hold onto a handle or handrail, or hands which shake so that holding items for any length of time is difficult; a lack of hands or fingers, or hands which are permanently clenched. All of these make holding, grasping, turning or pressing badly designed handrails, handles, switches, or buttons very difficult.

- **Visual impairments** – visual disorders can include total blindness or partial sight. Partial sight can involve various different visual experiences such as tunnel vision, patches of vision missing at the centre of the field of vision or at the edges; moving bodies floating across the field of vision; peripheral or side vision only; extreme lack of focus; flashing or flashes of light; loss of transparency; colour blindness involving perceiving some colours, often red and green, as mere shades of grey or brown. Other disorders while not strictly visual can affect the ability to understand what is being seen. This may be experienced by people with dyslexia and similar conditions, or for example with Asperger's Syndrome.

- **Hearing impairments** – this can include total deafness or hearing problems. Partial deafness ranges from mild to severe, and can involve sounds that are heard not only quieter, but distorted and speech which may be difficult to understand especially if there is background noise. There may be tinnitus, which is ringing or tapping sounds heard, or vertigo which results in dizziness and loss of balance. It may be that the sounds of speech are heard but become jumbled or distorted due to background noise, acoustic reflections, or electrical distortion in the auditory nerves.

- **Wheelchair use** – the reasons for using a wheelchair are varied. At one extreme, wheelchair users may be able to get out of their chair and walk for short distances or they may be able to get out of their chair if supported and move to a static chair or into a car. At the other extreme, they may be permanently seated into a chair specially moulded and formed to hold and support their body, perhaps with catheter bags, etc attached, and can only be lifted out by specially trained and experienced people, possibly with the help of lifting devices, and be unable to sit or be supported in any other type of seating device.

- **Size** – including unusual height i.e. very tall or small, unusually wide girth, children, pregnant women.

Other features are recognized by the DDA as defining disability such as mental impairments and incontinence. However, as stated, Part M of the Building Regulations is not about disability but is about allowing human beings to access and use a building. Therefore for the design of the building rather than the building in use and/or place of employment, which is covered by the DDA and other legislation, can concentrate on these six features of mobility, dexterity, vision, hearing, wheelchair use and size.

These six main features are not mentioned in Part M or the ADM because no definition or description is strictly necessary, but it will be helpful for the designer to bear them in mind while ensuring compliance with Part M, as these are the features concentrated on during the preparation of Part M and ADM.

We can look to these different features and note how poor design and detailing can create particular problems:

Mobility impairments – lack of suitable handrails; steep slopes or ramps; poorly designed steps and stairs; long paths or corridors with no resting facilities; poor arrangements in WC accommodation; desks and seating which are too low; sockets and switches set too low; lack of suitable sized parking spaces; trip hazards, gravel and other unsuitable paths; doors with closing forces which are too strong.

Dexterity impairments – poorly designed handrails, handles, and locking/security/opening devices; doors with too strong closing forces; poorly designed operating buttons for lifts, lighting, etc.

Visual impairments – poor contrast between wall and floors or walls and door frames or surrounds and sanitaryware; lack of manifestation on glazed wall and doors; trip hazards; lack of indication of change in levels; poorly sited switches/controls/sockets; poor lighting which produces areas which are too dark or result in glare; badly presented, unclear, or a proliferation of signage and information.

Hearing impairments – lack of hearing enhancement in reception areas, and rooms or spaces designed for performances/meetings/classes, etc; lack of visual information to support public address systems; lack of additional provision to support sound-based alarms; artificial lighting which is not compatible with electronic and radio frequency installations.

Wheelchair use – steps, stairs, poorly designed ramps, lack of turning circles, narrow doors, narrow corridors, thresholds, facilities placed too high (e.g. reception desks, vision panels, windows, screens, controls and switches), lack of accessible WC accommodation, gravel and other unsuitable paths, lack of dropped kerbs, lack of suitable sized parking spaces, doors with too strong closing forces. The spatial requirements of wheelchair users can often be the most exacting expected and wheelchair use is often therefore used as the benchmark for ensuring that people of all sizes and abilities can access a building.

Size – poorly designed and positioned handrails and handles; poorly sited opening and locking devices; poorly sited controls and switches; poorly designed sanitary accommodation; facilities placed too high (e.g. reception desks, vision panels, windows, screens, controls and switches).

It can be seen that some items of poor design can cause problems for different groups of people, so ensuring that particular items are well designed, provides better access for large numbers of people. It will also be noted that some of these items or issues are not covered by Part M and the ADM in their present form. However, they have been included above to give insight and to help the designer to choose and create better designs. This will help also in the longer term as the building is put into use, and perhaps comes under the control of the DDA which does cover these issues or items.

Anthropometric measurements have been provided in various publications to show the general space requirements that people need. The following measurements of required width of space, for example of a route, are typical, having been taken from empirical research:

- One person walking requires at least 0.55 m.
- A person using a stick requires at least 0.85 m.
- A child requires at least 0.5 m.
- One person pushing a pushchair requires 0.7 m.
- One person pushing a double buggy requires 1.0 m.
- One person pushing a buggy with a small child alongside requires 1.2 m.
- One person pushing a double buggy with a child alongside can require 1.5 m.
- One person using a wheelchair, or one person pushing another in a wheelchair, requires 0.85 m.
- One person in a wheelchair with a small child alongside requires 1.5 m.
- One person with crutches requires 1.2 m.
- One blind person with a cane (making sweeping movements) requires 1.2 m.
- One person in an electric wheelchair can require 0.8 m.

All of these figures of course are dependent on the size of the person, the size or shape of any aid they may use, and whether or not they require the presence of another person beside them, adult or child, or a dog.

When people pass each other, they require the above dimensions plus approximately an additional 100 mm. For example, a person with a stick passing a person using a wheelchair would need 0.85 m plus 0.85 m plus 0.1 m, or 1.8 m.

More space will be required for manoeuvering, turning, using the facilities, etc.

By considering of the above examples, one can see that for Part M 'Access to and use of buildings', designers need to take into account how much space real people will need, rather than too rigidly applying published figures as maxima, although these figures may be taken as minima. For instance, when providing a path to a public building, rather than ensuring that it is 1.0 m wide, designers should perhaps consider that there may be a parent and child walking along together, and design it as 1.5 m wide minimum clear width, with passing or waiting places. A footpath of 1.8 m would be much more pleasant and safe to use, particularly when bounded with high walls or hedges, and this width would be suitable for many people and combinations of people.

2.5 Interpretation – All buildings

The following definitions apply throughout Approved Document M:

Access – approach, entry or exit

Accessible – with respect to buildings, such that people, regardless of disability, age or gender are able to gain access

Closing force – see opening force

Contrast visually – the difference in light reflectance between two surfaces is greater than 30 points. This is used to indicate the visual perception of one element of the building or a fitting within a building against another. For example, a light switch plate compared to the surrounding wall, or a doorset against the surrounding wall. The greater the difference in light reflectance, the better the contrast enabling it to be more easily seen. Further information can be found in the Reading University publication, '*Colour, contrast and perception – Design guidance for internal built environments*' (University of Reading). The Light Reflectance Values are measured on a scale of 0–100, where 0 would be a perfectly absorbing surface and so totally black, and 100 a totally light-reflecting surface and so totally white. Different colours have different degrees of white and black or reflection properties. A value of 30 points difference between two adjacent surfaces gives a good contrast. A difference of 20 points may give an adequate contrast and could be argued to be so, but anything more similar and therefore lower than that is not adequate. There has not been enough research yet done on these, but current thought is that differences between two large surfaces is less critical than between a smaller one on a larger surface (e.g. a door handle on a door). More work needs to be done on the effect of surface textures, for instance the difference between glossy and matt surfaces.

Dwelling – a house or a flat. See also Student accommodation.

Independent access – access to a part of a building (from outside, and therefore from the site boundary and from any car park on site) that does not pass through the rest of the building.

Level – being predominantly level but having a maximum gradient, along the direction of travel of 1 in 60. This applies to surfaces of a level approach, access routes, and landings to steps, stairs and ramps.

Opening force – this is the effort necessary to open the door from the closed position. It is harder to open it from closed through the first 30° or so. It is measured in Newtons and can be measured using a plunger-type force measuring instrument or a spring-loaded balance, hooked onto the door handle that a person has to use to open the door. The accuracy of these varies. The force in most situations should not exceed 20 N. For double swing doors, the method of opening in both directions is by pushing. It is much easier to push a door to open it than to pull it, and so an opening force of 30 N is usually recommended. Sometimes there are conflicts with the closing force needed to hold a fire door closed, in which case an alternative provision for closing the door, should be used. A fire door with too strong a closing force would not be acceptable for safe means of escape. By careful selection of components, door closers can be specified and installed that will meet the requirements of both Part M and Part B of the Building Regulations. Doors with closing devices need to be regularly maintained to ensure both the closing and opening forces are adequate and not outside acceptable limits.

Principal entrance – the entrance which a visitor not familiar with the building would normally be expected to approach.

Student accommodation – this type of building is treated differently to flats. Where new blocks of flats is constructed for student accommodation, the space requirements of the sleeping accommodation and the facilities (WC provision, switches, etc.) are designed as though they were hotel/motel accommodation, which is in the section on 'Buildings other than dwellings' in Approved Document M, while the rest of the building should be designed in accordance with the requirements for flats, in the section on 'Dwellings' in Approved Document M. There are reasons why this should be so from the viewpoint of both the occupier and supplier of the accommodation. In a normal rental situation, the potential occupier has a choice as to whether to take the offered accommodation if it meets their general needs. A student taking accommodation for the first time, possibly with minimal notice of their attendance at that place of higher education, may have little choice. Accommodation built to meet the general requirements of people with a range of abilities (as is the case for non-dwelling buildings) will be of more use in this situation than the limited provisions required for dwellings.

Suitable – designed for use by people – regardless of ability, disability, age or gender. In Part M, this is in relation to means of access and use of facilities, but is subject to the usual gender-related conventions regarding sanitary accommodation.

Usable – convenient for independent use. This is with respect to buildings or parts of buildings, independent use implying that a person would not need another's assistance to be able to use the building or part in question.

2.6 Interpretation – Dwellings only

Certain other definitions only apply to dwellings as given below. These are also given at the start of the chapter on dwellings.

Clear opening width – For dwellings, the clear opening width of a door is taken from the face of the door stop on the latch side to the face of the door when open at 90°.

Common – serving more than one dwelling.

Dwellings – This term means a house, flat or maisonette. It does not include hotel accommodation or motels; this type of accommodation is covered under 'Buildings other than dwellings'. Purpose built flats used as student accommodation are regarded as a mixture of both, with general provisions as 'Dwellings', but in respect of space requirements and internal facilities they are to be treated as hotel/motel accommodation in 'Buildings other than dwellings'. One reason for this is that, similar to hotels, the provider of the accommodation may have little knowledge of the requirements of the occupier, and more particularly the occupier may have little choice in their accommodation. When renting or buying a house or flat, however, the occupier usually has the choice to accept the accommodation or reject it if it does not meet their needs. Therefore, refer to 4.6.7 onward for advice regarding space requirements and internal facilities for student accommodation.

Entrance Storey – This is defined in the Regulations for requirement M4 (dwellings only) as meaning the storey which contains the principal entrance.

Habitable room – This term is used for defining the principal storey of the dwelling. It means a room used, or intended to be used, for dwelling purposes and includes a kitchen, but not a bathroom or utility room.

Maisonette – a self-contained dwelling, but not a dwelling house, which occupies more than one storey in a building.

Point of access – The point at which a person visiting a dwelling would normally alight from a vehicle, which may be inside or outside the boundary of the premises, prior to approaching the dwelling. For instance, a visitor may arrive by car. This car may be able to come up the drive and the visitor can alight near the front door of the dwelling. Alternatively, they may have to be dropped off at the kerb outside and make their way up the drive or garden path. The point of access will be wherever the provision is made for the visitor to alight.

Principal entrance – The entrance which a visitor, not familiar with the dwelling, would normally expect to approach, or the common entrance to a block of flats.

Principal storey – This is defined in the Regulations for requirement M4 (dwellings only) as meaning the storey nearest to the entrance storey which contains a habitable room, or if there are two such storeys equally near, either such storey. For example, one may enter the dwelling by the front door on a level 'A' which contains only an entrance hall and the stairs to an upper level 'B' where, say, the kitchen and dining room are. Level A is the entrance storey, and level B is the principal storey. Where the entrance storey contains the habitable rooms, the principal storey and the entrance storey would be the same.

Plot gradient – The gradient measured between the finished floor level of the dwelling and the point of access.

Steeply sloping plot – A plot gradient of more than 1 in 15.

Further definitions taken from BS8300 are given in appendices.

2.7 Application

Part M of Schedule 1 to the Building Regulations 2000 applies to:

(a) a newly erected non-domestic building;
(b) a newly erected dwelling;
(c) a new extension to an existing non-domestic building;
(d) a material alteration to a non-domestic building;
(e) a material change of use to an existing building to form a hotel or boarding house, an institution, a public building or a shop.

The requirements for Part M do not apply to an extension or material alteration of a dwelling. However, under Regulation 4(2), where any building, including a dwelling, is extended or undergoes a material alteration, the work carried out must not make the building any more unsatisfactory than it was before. Therefore, if a dwelling previously complied with parts of Part M, it must not be extended or altered in such a way that it no longer complies.

The requirements for Part M also do not apply to any part of a building which is used solely to enable the building or any service or fitting in the building to be inspected, repaired, or maintained. This would be reasonable, as ordinary people would not expect to be able to access these areas. Other than these, all areas should be accessible.

2.8 Extensions and alterations

When an existing building is being altered, the alteration should be considered to determine whether or not it is a material alteration. A material alteration is work which:

- would result in the building not complying with Part M where previously it did; or
- if the building previously did not comply with Part M, it would result in the building being more unsatisfactory in relation to Part M.

So a planned alteration or extension that has the potential to reduce the compliance of the building as a whole with Part M, must be carried out in such a way that there is no lessening of the current level of compliance.

Additional works may therefore be necessary. These works are however limited to the works to the building itself. The Approved Document suggests that it is not necessary for a material alteration for Building Regulations to upgrade the access from the site boundary or car park to the building's entrance provided it is not made worse, although this may be sensible and advisable, for example under the DDA.

When an existing non-domestic building is being extended, the actual extension should be treated as though it is a new build, and should comply fully with Part M. The existing part need not be brought up to acceptable standards. To make this sensible, where possible, the extension should have a suitable independent access; that is, access to the extension (from outside, and therefore from the site boundary and from any car park on site) that does not pass through the rest of the building. This will ensure that people can access the new part of the built environment even if the existing part is not accessible. It will negate any requirement to upgrade existing sub-standard property, at least under this legislation. If the owners do not wish to provide independent access then they can show that suitable access to the extension can be gained through the rest of the building. This may mean modifying the existing building and the ADM claims that this work would then be a material alteration, which will be subject to Regulation M1.

On reading the definitions of material alteration in Regulation 3 of the Building Regulations, it is not easily apparent that this modification would be a material alteration because the tests are that (a) the building does not comply when previously it did, or (b) the building did not comply and now it is worse. A modification to provide a suitable access would not fall into either of these camps. The way to check whether a modification is a material alteration or not is to consider that if the modification is carried out so that it does not comply with a particular requirement (A, B, or M), and this makes the building or part not comply when it should, the modification actually jeopardizes the building or part. The modification therefore has to be a material alteration, which complies, or the building will be in contravention.

For example, shown in Figure 2.1 is existing Building, B, which has two existing doors, d_1 and d_2 of clear width 790 mm each. Building B is to have an extension, E. No external doors are to be provided in E but an access is to be made into B via a new door d_3, of clear width 1000 mm, and a new external door, d_4, is to be provided.

If door, d_4, is formed to be of the same (unsuitable) width as the existing doors, d_1 and d_2, then there is still no suitable access to the extension gained through the rest of the building and the extension does not comply with Regulations M1 or M2. So the formation of d_4 has to be a material alteration in order for the building's extension to comply.

Similar to domestic property, an extension or material alteration to an existing non-domestic building should not make the existing part less accessible than it was before.

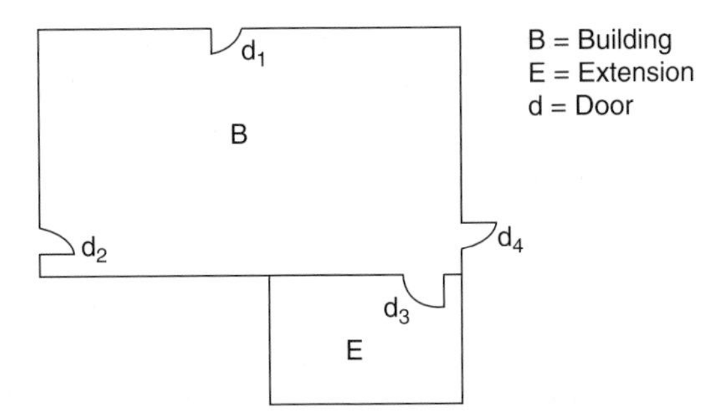

Figure 2.1 Example of access to an extension

Example case study 1

A hotel owner extends the premises to create a new conference room to the rear adjacent to the car park. Independent access is provided as well as access through the hotel. The WCs are positioned further into the existing hotel. Reasonable provision will need to be made for people to gain access to the WCs from the conference room. This may involve increased structural alterations.

2.9 Change of use

Where there is a material change of use, Part M may be applicable. When the new use is to be a:

- hotel
- boarding house
- institution
- public building
- shop

then the building must be upgraded where necessary to comply with requirement M1, which states that reasonable provision shall be made for people to gain access to and use the building and its facilities. Definitions of the above uses need to be taken from the Building Regulations 2000, regulation 2. Example uses are shown in Table 2.1, which also includes further clarifying examples taken from Approved Document M.

If only part of the building changes its use, then for any of the uses shown in Table 2.1, regulation M1 applies to that part of the building. In addition, this includes any sanitary conveniences provided in connection with that part.

Table 2.1 Examples of use for change of use requirements

Use	Includes for example
Hotel or boarding house	Motel, residential college, hall of residence, hostel
Institution	E.g., hospital, home, school, nursery, etc. used as living accommodation for, or for treatment, care or maintenance of, people who sleep on the premises, who are: Suffering from disabilities due to illness, old age, or other physical or mental incapacity Or are under the age of 5 years
Public building	Theatre, public library, public hall, etc. School or other educational establishment Place of worship Non-residential day centres, clinics, health centres and surgeries But not places where members of the public are only occasionally admitted
Shop	Restaurant, bar, public house, café, etc Places used for sale of food or drink to members of the public Places used for retail sale by auction Barbers or hairdressing premises Premises used by the public for hiring any item Premises where the public take goods for repair or other treatment

This means that if, for example, a multi-used or occupied building has common parts which include the WC accommodation, this WC accommodation is likely to need upgrading where there is a change of use of any part to one of the above uses.

Example case study 2

The owner of a building used for offices converts the ground floor office to a coffee shop. The toilets for the offices are on the first floor. The coffee shop will need toilets for public use. Reasonable provisions will need to be made for people to gain access to and use these facilities. Alternatively more (accessible) sanitary conveniences could be created inside the coffee shop.

Example case study 3

The owner of a building used for offices of a manufacturing firm converts the first floor offices to a showroom where members of the public can view and order the goods. Reasonable provisions will need to be made for people to gain access to this showroom. Suitable independent access will need to be provided or suitable access through the building to that part.

2.10 Historic buildings

Historic buildings can include a number of different categories of building, for example:

- listed buildings
- ancient monuments
- buildings in a conservation area
- buildings of historic or architectural interest which are in national parks, world heritage sites, areas of outstanding natural beauty, and so on.

Historic or even listed buildings are treated no differently to any other in relation to Part M or the DDA. That they have special features which are desired to be retained does not mean that they do not need to comply with Building Regulations, including Part M. Conversely, access measures do not override the need to obtain, for the proposed building works, planning permission, listed building consent or English Heritage acceptance; and compliance with Part M, which requires 'reasonable' access, does not mean slavish adherence to Approved Document M if this would result in the destruction of architectural or historically important features.

There is no statutory or other duty on planning authorities to have special regard for disability issues, which is not to say they don't – in the same way that they consider fire safety and other health and safety issues. PPG15 states in paragraph 3.28, " It is important in principle that disabled people should have dignified easy access to and within historic buildings. If it is treated as part of an integrated review of access requirements for all visitors or users, and a flexible and pragmatic approach is taken, it should normally be possible to plan suitable access for disabled people without compromising a building's special interest. Alternative routes or re-organizing the use of spaces may achieve the desired result without the need for damaging alterations".

English Heritage have produced an excellent guide with photographs and examples, available on the web, entitled, Easy access to historic properties. (English Heritage, 2004). English Heritage has a very positive attitude to access and historic buildings; they regard themselves as the lead advisory body in providing access to historic buildings in England, and state, "We believe access should be celebrated with high-quality design that is also sensitive to the special interest of historic buildings." (English Heritage, 2004).

What is important to planning and listed building officials as well as English Heritage is the detailed design. Relatively minor features can make or break a design architecturally, just as they can from an accessibility viewpoint. For example, handrail design and door design can be designed well, or provided without thought, for both or either aspect. As the performance standards of construction become more understood and expectations heightened, so should the detail design be given more consideration. This holds true not just for historic/architectural interest reasons, but for other aspects such as avoiding thermal bridges and interstitial condensation, improving sound insulation and maintaining good internal air quality.

If a designer is considering works to improve access to an historic building, it would probably be worthwhile consulting the local authority planning and listed building departments before a planning application is finalized in order to gauge the likely requirements. The benefit is twofold in that a number of differing solutions can often be identified during the meeting, and also the Planning officials become more aware of the constraints and needs of a building's occupants. Once planning and Listed Building consent is gained, these considerations can then be incorporated into the Building Regulation submission.

Great use should be made of the Access Statement to accompany a Building Regulation submission, to clarify the constraints posed by the existing building, its environs and the historic importance of each. Where full access is unreasonable or excessively costly due to these, compensatory measures can be proposed in the Statement which may not be building works related.

3

Access statements

3.1 General

The use of Access Statements has been introduced into the latest edition of the Approved Document to Part M. There is the ability to depart from the usual design guidance in a clear and transparent way by use of an Access Statement submitted with the Building Regulation application. This has been introduced into ADM to enable the designer to demonstrate that the designs and details proposed fulfil the same standard or level of performance and functional criteria as that guidance given in the Approved Document. This is not an easy option, or way of just not conforming to the Approved Document, because, in order to depart from the approved guidance in a satisfactory, reasonable and sensible manner, one first has to understand and appreciate what the approved guidance says and why it says that.

If the proposal complies fully with ADM, there is not a requirement to submit an access statement. However this is perhaps a lost opportunity for designers as an access statement can assist the design team to focus on their design and level of inclusivity, form a useful tool for the client in the property's health and safety file (under the Construction (Design and Management) Regulations), and assist the occupier in duties under the Disability Discrimination Act (DDA).

Access Statements are also required for planning applications. The concept of an Access Statement has been introduced into guidance for planning and for building control applications as a means to demonstrate that work, for example new build, extensions, repairs, replacements and refurbishment, is designed to address the obligations of reasonableness contained in the DDA 1995. By producing and developing a document, which is intended to pass to those who will undertake the long-term management of the building, the Access Statement is designed to assist in ensuring that the 'evolving duty' placed on service providers, employers and educators under the DDA can be better addressed (DRC, 2004).

The Disability Rights Commission in the guidance given for Access Statements state that a correctly developed Access Statement will provide an opportunity for developers, designers, product providers and managers of environments to demonstrate their commitment to ensuring accessibility in the work they undertake. It will allow them to demonstrate how they are meeting, or will meet, the various obligations placed on them by legislation, and how they will continue to manage accessibility throughout the delivery of the services they provide or the employment opportunities they create.

Approved Document M recommends that an Access Statement is deposited at the same time as (1) plans are deposited for the building regulation application, (2) the building notice, or (3) the details are given to an approved inspector, dependent on the method chosen. It recommends that the Statement be updated to reflect decisions on site, which would include decisions made before getting to site. Of course, the interesting situation remains that Access Statements (Building control) are only a recommendation contained in the Approved Document, and not a requirement of the Building Regulations. However, as transparency makes sense on a number of levels, the authors of this book recommend that full use is made of Statements.

Access statements were introduced for the planning process (rather than building control purposes) by the Office of the Deputy Prime Minister (ODPM) in 2003 in the publication 'Planning and Access for Disabled People – A Good Practice Guide'.

They have now also been brought into recommendations for building control. The two documents unfortunately have been given the same name; they are not the same document, but neither are they entirely separate. The Access Statement (Building control) is expected to be complementary to, and developed from, the Access Statement (Planning). Note that this book, not the ODPM, has given the bracketed distinguishments.

The requirements for each are discussed below.

3.1.1 Access statement (planning)

Access statements are to be required for planning permission in accordance with Section 42 of the Planning and Compulsory Purchase Act 2004.

'Planning and Access for Disabled People – A good practice guide' suggests that an access statement be provided in order to identify the philosophy and approach to inclusive design, which has been proposed in the proposal put in for planning consent. It requires the identification of key issues of the scheme, and the sources of advice and guidance used to determine the design.

Because of the nature of planning applications, the considerations at this stage may be very 'broad-brush'. The detail has not yet been determined, but the intention is there. The challenges, problems, compromises, etc. expected are highlighted, as well as the items where 'compliance' will be a matter of course.

Then, in a similar manner to the pre-tender health and safety plan for the CDM regulations being taken up and expanded on in detail by the contractor to form the post-contract health and safety plan, the Access Statement (Planning) is taken up and expanded in detail to become the Access Statement (Building control). This is a simplistic illustration because the same designers are likely to do both statements whereas, of course, for the CDM regulations the contractor develops the plan.

3.1.2 Access statement (building control)

To show compliance with Building Regulations, the detail has to be worked out. This is the same whichever part of the regulations is being considered. As with other parts, the precise system or material may not be known at this stage, but the designer should know the pertinent or generic qualities they require. There is likely to be a number of ways that compliance with the regulations could be met.

Regulation M1 requires that:

Reasonable provision shall be made for people to gain access to, and use, the building and its facilities.

There is a lot of scope and room for interpretation in that. There is a lot of guidance published, which may be of equal relevance to, if not better than ADM, in the particular circumstances.

If the designer chooses not to conform to ADM, it is in everyone's interests to show how the building is to comply with M1, and why this is a good idea. If the designer chooses to conform with ADM, then in the Access Statement (Building control) they should state when, and where, the building does so.

If the designer chooses to comply by a different method, the Access statement should state why, what the standard of references used are, and describe the compromises and priorities that have had to be made to ensure that the most reasonable access is being provided given the specific circumstances.

For access issues it is not obvious that compensatory features can be made if one cannot reasonably provide access to a particular facility. In Part B, for fire safety, the designer may wish to reduce the level of fire resistance of certain elements of structure, and to compensate, may provide a sprinkler system for instance. In these circumstances, the sprinkler system may well compensate by not allowing a fire to develop at such a fast rate and therefore enables occupants to make their escape before conditions become untenable. On the other hand, however, the provision of a hearing loop is unlikely to compensate a wheelchair user for the lack of an accessible WC.

Notwithstanding that, it may, for example, not be reasonably possible to alter an existing small building to make the one WC fully accessible, but it may be perfectly reasonable to provide, in the existing sanitary accommodation such items as grab rails, a shelf for colostomy bags, easily worked opening and locking devices set at a good height, good colour contrast and an alarm pulley. All these and other provisions may make the WC much more accessible and usable for many disabled people. This is a compromise of course, but one that should be highlighted as a good improvement for many people.

If the access statement is made as full and detailed as possible, the building control bodies will be able to see clearly and easily the issues having to be dealt with due to the configuration of the building or the site, its use and/or users, reasonable limitations such as its listing or monetary factors, or other factors, which the designer has had to work with. By stating the overall and specific philosophy of access and the different standards or publications worked to decide a solution, they will therefore be able to be more understanding of the scheme and if necessary may be better able to suggest other or different measures, which may be of benefit to it. ADM therefore recommends the provision of Access Statements. Examples of Access statements can be seen on the DRC website. www.drc-gb.org.

3.1.3 Format of the access statement (building control)

The format of the Access Statement is not standardised because issues such as the size, use, nature and complexity of the building and the proposed work will affect its access requirements. However, the Statement should address the following:

- The policy and approach to access being adopted, including reference to the inclusion of disabled people.
- Any specific issues affecting access to the particular building, the solutions chosen to overcome these, and the reasons for this choice.
- Sources of advice and technical guidance followed, including any user consultation planned and undertaken.
- The access strategy being implemented, which would include management policies. This would include an explanation of specific issues and the design solutions adopted to overcome these issues.
- Any action plan detailing the implementation of the access strategy.
- A plan of the building and site illustrating entrance and exit routes and circulation around the site and the building, indicating any specific or relevant features.
- Where good practice cannot be achieved, the Access statement should explain this and state what mitigation measures are to be implemented.

The Access Statement should state the guidance publications and reliable standards that have been used in the design. Where the accepted 'normal' guidance is impractical or unreasonable in the specific circumstances, this should be explained and the reasons for adopting alternate standard or guidance clearly stated. In the event of any legal challenge, the access statement may be called on for evidence and should therefore be robust. The justification for design decisions should be strong, and the record of events clear.

3.1.4 Continuity

The Access Statement is intended to be a living document. This is not easy to achieve. The DRC publication on Access Statements (DRC, 2004) sets out a four - stage strategy:

(1) The Strategic Access Statement:

This should be at the project brief stage, expressing the level to which accessibility is considered by the management or owner.

(2) The Access Statement at Planning:

Building on the Strategic Statement, this would be used for the planning application and include:

- Details of the site plan and positioning of buildings on the site
- Access issues considered in formulating the planning application
- Planning guidance and legislation used
- Any consultations which have taken place
- Any professional advice taken
- The suitability of public and other transport links
- Any overview assessments of technical requirements or decisions, *e.g.* means of escape, auxiliary measures, information and communication strategies, etc.

(3) The Access Statement at Design:

Building on the previous two statements, this stage of the Access Statement will provide a more detailed description and explanation of the design. It will cover:

- The philosophy of the design team in terms of accessibility
- Which guidance and legislative standards have been adopted and why
- Details of design issues which deviate from established practice
- Constraints of the building, *e.g.* in existing buildings
- Mitigation
- Stage 1 Access Statements from suppliers of services and equipment *e.g.* for the provision of audible and visual fire alarm systems
- How the Statement will be developed for use by the building managers, occupiers, etc.

(4) The Occupancy Access Statement:

This will draw on all the previous stages of the Statement and will demonstrate the commitment to the measures provided to enhance accessibility. This will involve:

- An assessment of the measure provided to ensure their suitability
- Policies to ensure the appropriate maintenance of provisions *e.g.* internal and external pedestrian routes
- Policies to introduce measures recommended in stage 2 or stage 3 Statements, *e.g.* management policies, training of staff, decoration schemes, etc.
- Policies to ensure the on-going suitability of measures introduced or recommended in stages 2 and 3 Statements, *e.g.* the need to ensure on-going colour and luminance contrast in future changes to colour schemes, the need to maintain lighting levels, on-going appraisal of information provision
- Details of suitable management approach to on-going maintenance of essential features, such as lifts and induction loops, and evidence of the management practices and policies that have been put into place to ensure appropriate prioritisation of repairs and maintenance

Much of these should find a home in management tools such as planned maintenance policies and facilities management policies. This stage 4 Access Statement will be a major tool for service providers and others regarding their duties under the DDA.

It is not expected that these four stages will create a formal document in all cases of new work. The size, complexity and even existence of each stage statement will depend on the size, complexity and nature of the work proposed. However it is likely that each stage will be at least informally considered.

3.1.5 Example access statements (building control)

The following is an example of an access statement to accompany a building regulation application to alter and refurbish a small commercial property. It is adapted from an example given by the DRC (2004).

Proposed development

The client intends to fit out a shop to form a coffee bar. The premises consists of a ground floor retail area, basement and first floor. There is a second floor, but this is to be sealed closed and not used.

Access Statement for 12 Coffee Court

Proposed Use of the Development

Fit out of an existing ground floor/basement retail unit to form a coffee shop.

Description

An existing retail unit is to be converted to a coffee shop. There is a 150 mm step at the ground floor entrance and access into the basement area, which houses storage areas and toilets for staff, is via a single staircase. It is proposed to fit out the ground floor (120 m^2) and the basement area (84 m^2). The basement will remain mainly as storage, although it is proposed to provide two customers toilets in addition to the staff toilet accommodation.

Access Statement – Stage Two and/or Three

Access Statement for Bean's Coffee Shop, 12 Coffee Court

Background:

Bean's & Co. has recently acquired the lease to the above premises and proposes to upgrade the accommodation and re-designate its use to a coffee shop. Bean's Ltd. is a small retailer with only two existing outlets, this being the third.

Bean's & Co. propose to fit out the ground floor area as a coffee area furnished with loose tables and chairs (48 seats and 16 tables), a server and a display area. The basement area, which can be accessed only by stairs, will be refurbished and will remain as staff/storage accommodation, although it is proposed to install two new toilet facilities for customers. The first floor will not be altered and be used for office accommodation. There will be no public access to the first floor.

Approach/Entrance:
The proposal is for the existing shop front to remain. However, following consultation with the Highway Authority, agreement has been reached to raise and re-grade the footpath outside the unit to eliminate the obstacle of the single 150 mm step at the entrance.

To provide a level area outside the entrance door, the door will be recessed and the original doors reused. The doors provide a minimum clear opening width of a single leaf of 800 mm and the doors will be maintained to keep the required opening force to a maximum of 20 N.

Existing door furniture will be replaced with pull handles commencing at 800 mm above finished floor level and which contrast in colour and luminance with the door.

Circulation:
The ground floor is level throughout and the existing timber boarding will be covered with slip resistant vinyl.

The basement has single steps into each room. The ceilings are low at 2.0 m above the floor level to the corridor. It is not proposed to alter the steps into the storage areas. To avoid risk of trips for the public, the new sanitary accommodation will have a raised floor formed by concrete incorporating damp proofing measures. This will reduce the room height in the new sanitary accommodation, but is a compromise measure.

Counter:
The counter/servery/display unit being installed is 1200 mm high. This height is governed by the Health and Safety requirements relating to the location/use of standard coffee machines.

A table service/assistance will be made available to all customers who are unable to safely carry/collect goods from the counter and transport them to the tables.

Loose furniture will be moved by staff as necessary to maintain clear and safe access routes.

Staff Accommodation:
The existing staff accommodation, including staff room and toilet is located at basement level and is accessed via a single staircase only. The provision

within the staff area currently meets the needs of all employees who have been appointed to work in the unit.

Should a disabled person be employed at the store, Bean's & Co. has a stated policy for identifying individual needs and undertaking reasonable adjustments, as identified in Part II of the DDA 1995.

Customer Toilets:

There are currently no customer toilets available in the unit.

It is proposed to improve toilet facilities by providing customer toilets at basement level. The accommodation will comprise two cubicles, one being designed and fitted out in accordance with the recommendations in ADM 2004 edition for use by people with ambulant disabilities.

The provision of an accessible toilet at ground floor level has been considered. However, providing this facility would result in the loss of four tables (16 seats), reducing considerably the viability of the premises.

The provision of lift access from ground to basement level was also considered. This would result in a similar loss of tables, plus the additional financial cost of the necessary structural alterations and purchase of equipment.

The existing staircase is to be upgraded by providing an additional handrail to the design recommendations of ADM 2004 edition, with appropriate colour and luminance contrast being provided to the handrail and step nosings.

Lighting will be improved to meet the recommendations of the Chartered Institute of Building Services Engineers (CIBSE) Code for Lighting.

A notice advising of the WC facilities available in the premises will be provided adjacent to the entrance and on the Bean's & Co. web site.

Fire safety:

There is an exit to the rear of the premises at ground floor level to the back yard, which gives access to a back alley.

This door is kept unlocked during opening hours. It has a clear opening width of 800 mm.

There are two steps of 150 mm rise and 250 mm going down into the yard. It is proposed to provide a handrail to the design given in ADM 2004 edition.

Due to the size of the ground floor it is unlikely that means of escape will need to be afforded via the rear yard for customers. However, the shop manager and assistant manager have manual handling training, which includes guiding a wheelchair down these two steps.

The gate to the back alley is kept locked. It is proposed to alter the locking device on the gate to one, which can be used in the case of an emergency without the use of a key during opening hours.

A new break glass alarm system is to be incorporated with sounders for the basement storerooms, basement sanitary accommodation, and first floor offices. A visual signal will also be incorporated to the sounder in the sanitary accommodation.

The second example involves a new industrial unit, which has Building Regulation approval. The prospective purchaser of the unit needs further works inside the unit in the form of a two storey office and plans are being put in for the provision and fit out of this office.

Access Statement for Workshops United Ltd.

To be read in conjunction with architect's plans of fit-out.

Proposed development
Workshops United Ltd. intend to fit out a light industrial unit including incorporating a two-storey office unit. The premises consist of a ground floor workshop area, ground floor offices, staff room and toilets and first floor offices.

There will only be commercial deliveries/visitors to the premises, and no general public. No unaccompanied visitors will continue beyond reception.

Description
The fit-out of the industrial unit involves creating, inside the unit, two-storey offices, each floor plan of 19 m $\times$ 11.5 m.

Parking
One parking bay designated and suitable for disabled people is provided close to the principal entrance. This measures 2400 mm x 4800 mm with an access zone to one side of 1200 mm width and to the end of 1200 mm width, as shown on plan detail.

Approach
The approach from the highway will remain to the industrial unit as existing approved. There is level access from the car parking area to the building entrances with any kerb provided with dropped kerbs as shown on plan detail.

Entrance
Finished floor level is approximately 200 mm above external ground level. A ramp will be formed to the main entrance and a level threshold provided, as shown on plan detail.

The doors provide a minimum clear opening width of a single leaf of 900 mm and the doors will be maintained to keep the required opening force to a maximum of 20 N.

Door furniture will be of a type easy to operate by people with limited manual dexterity, positioned at around 1000 mm above finished floor level and contrast in colour and luminance with the door.

Circulation

The ground floor is level throughout and covered with slip resistant floor covering.

Internal doors provide a minimum clear opening width of a single leaf of 900 mm and the doors will be maintained to keep the required opening force to a maximum of 20 N.

Staff accommodation

The staff accommodation, including staff room and toilets are located on the ground floor. The provision within the staff area currently meets the needs of all employees who have been appointed to work in the unit.

Should an employee have or develop a disability, Workshops United Ltd. has a stated policy for identifying individual needs and undertaking reasonable adjustments, as identified in Part II of the Disability Discrimination Act 1995.

Accessible sanitary accommodation

A unisex accessible WC is provided adjacent to reception and the main entrance door at ground level. The floor area provided is 2200 mm × 1600 mm. The internal layout and provisions will be as shown on plan detail.

The main staff toilets include a male and a female WC designed for ambulant disabled people. These have a floor space of 3600 mm × 1200 mm. Both include a door between the WC and washbasin areas; this partition and door will be so positioned that a clear space of 750 mm length remains in front of the WC pan, unobstructed by the door swing, as shown on plan detail.

Vertical circulation

The stairway is in compliance with the provisions of Part K and Approved Document K. The risers are 170 mm maximum, the goings minimum 250 mm. Handrails are provided on both sides of the flights at a height of 900 mm above the pitchline and 1100 mm above landings.

The office unit is relatively small, there are no unique facilities to the first floor and the proposed use is for no public access to the premises. As such it would not be reasonable to require a passenger lift. It is therefore proposed that an enclosed lifting platform is provided to the premises to provide access to the first floor for persons of impaired mobility. The platform will be incorporated in the area designated as store on the proposed plans.

- The platform will be of an enclosed type conforming to the Supply of Machinery (Safety) Regulations 1992, SI 1992/3073.
- The platform will travel more than 2 m vertically but less than 3 m.
- The rated speed will be less than 0.15 m/s.
- The control positioned between 800 and 1100 mm above the platform floor and at least 400 mm from a return wall.
- Continuous pressure controls will be provided.
- Landing call buttons will be between 900 and 1100 mm from the ground and first floor landing floors.
- The minimum clear dimensions of the platform will be 900 mm wide and 1400 mm deep, which would cater for an unaccompanied wheelchair user.
- The doors will have a clear effective width of at least 800 mm.

Fire safety:
Means of escape provision is in accordance with Part B and Approved Document B.

The lifting platform will not be of a standard for means of escape. The area in front of the lifting platform will be designated as the refuge area for any wheelchair user.

There will be no unaccompanied visitor access to the first floor and Workshops United's management procedures and fire risk assessment measures will identify the needs to be taken for means of escape for specific staff members.

A fire alarm system complying with BS5839-1:1998 and an emergency lighting system in accordance with BS5266-1:1998 are to be installed, along with required extinguishers, notices and signs.

4

Buildings other than dwellings

4.1 Introduction

4.1.1 The main provisions

Provision should be made for suitable access to, and into buildings including the use of their facilities. In order to achieve access to buildings other than dwellings, consideration should be made for:

- A suitable approach from the boundary of the site to the principal entrance
- Access into the building
- Access within the building
- The use of facilities, including sanitary accommodation
- Means of escape.

4.1.2 Interpretation

Definitions given by the Approved Document M (ADM) are listed in section 2.5. There are no additional definitions specific to non-dwellings.

BS8300 also gives definitions. For a comparison see Appendix 1.

There is little difference of interpretation between the two documents, where they both define a term. The main difference is for the word "access". ADM defines this as '*Approach, entry or exit*' whilst BS8300 states '*Access to, and use of, facilities and egress except in cases of emergency*'. BS8300 seems to suggest that means of egress in cases of emergency are not the subject of this British Standard. Indeed there is little mention of means of escape in BS 8300.

The Approved Document seems to suggest, in its shorter definition, that as means of escape is not excepted, it is included. However this is not the case. ADM states in a note to the section on the Requirements of the Building Regulations that '*the scope of Part M and ADM is limited to matters of access to, into, and use of a building. It does not extend to means of escape in the event of a fire for which reference should be made to Approved Document B*'.

This should be borne in mind if designing to either the British Standard or ADM and reference made to other guidance such as Approved Document B, Fire Safety, and the BS5588 series: Fire precautions in the design, construction and use of buildings. If during the design process, the designer considers disabled people specifically, then BS5588-8:1999, the Code of Practice for means of escape for disabled people, should be consulted. This Standard is also referred to in Approved Documents B and M. Appropriate systems for alerting occupants, who may have hearing, sight and/or learning difficulties should also be included. Further discussion of means of escape is given at 6.4.4.

4.1.3 Objectives of the provisions

For buildings other than dwellings, the objectives of the requirements are for people to be able to gain access to the building and use the facilities of the building. To be able to do this, the aim is to achieve a good means of access for people from the entrance point on the site boundary, and from any car parking provided within the site. People may arrive at the building under their own efforts or by car and therefore provision should be made for both.

A good means of access would be a route that does not incorporate obstacles and hazards, and is not problematic to people, including those who have impaired sight, impaired mobility, use walking aids, or use wheelchairs. Hazards en route could include windows that in open position present edges at child height.

Some solutions to obstacles may not be useful to different people. For example, a change in level may be overcome by a ramp, but this can be uncomfortable for people with impaired movement in their ankles; such people may prefer a stair. Where possible, in this situation, a stair and ramp could both be provided, or a compromise may be the better solution in the circumstances.

On reaching the building, people should be able to access it via the main or normal entrances without difficulty, and then move around the building without encountering obstacles.

For new buildings, the aim is that the principal entrance and the staff entrance should be accessible. Where an existing building is being extended, ideally the main entrances would be accessible to access the extension, but where this is not possible an alternative provided which does give access would be acceptable.

4.2 Getting to and into the building

4.2.1 Approaching the building

Access to the principal entrance, and any other entrances, should be provided from entry into the site curtilage and car parking. Access from one building to another on the site should also be considered.

Where possible, the route should be level between the site boundary and any accessible parking to the principal entrance. In some cases this is not possible however access should be provided to alternative accessible entrances. It should be borne in mind that under the Disability Discrimination Act 1995 (DDA 1995) this alternative entrance should not discriminate any disabled user in terms of the service that is being provided. In circumstances where the gradient of the approach is 1:20 or steeper, alternative provisions should be made (i.e. ramps, lifts and steps).

ADM recommends pathways to be of surface width 1800 mm, however in restricted sites 1200 mm should be sufficient where justified in the Access Statement. In circumstances where 1200 mm pathways are present, passing places should be included that are 1800 mm wide and 2000 mm long at intervals of no more than 50 m.

Pathways should be firm, durable and slip resistant and avoid difference in levels at joints in the paving. Pathways should, where possible, be free from obstruction such as bins and seating, and overgrowing vegetation.

Tactile paving should be provided on pedestrian routes for visually impaired people, where there is an uncontrolled vehicular crossing point. Tactile paving is generally of two types – blister or ribbed. The blister type is used at crossing points of roads, indicating danger in the form of vehicles. The ribbed or 'corduroy' type is used as an advanced warning of a change of level for example at the bottom and top of flights of steps.

Tactile paving is an example of conflicting interests, in that many people with visual disorders find tactile paving extremely helpful to the point of essential, while some people with mobility problems and sensitive feet can find them a trip hazardous and very painful to walk across. Other people who are using items such as shopping trolleys, pushchairs, wheeled luggage etc. also find both types obstructing.

To avoid the worst problems, the raised ribs should be no higher than 6 mm above the paving level and blisters no higher than 5 mm (see also Figure 4.1).

4.2.2 Car park provision

For people who arrive at the building by vehicle, access to the property needs to be available from the place where they leave that vehicle. In ADM this is considered when this point is at an on-site car park space or setting down position.

People need adequate space to leave the vehicle, to move to the rear of the vehicle, if required for example to retrieve a mobility device, to buy a parking ticket if necessary, then to move to the principal or alternative entrance. Barriers and obstacles to this movement should therefore be avoided.

Consideration should therefore be given to the following:

- The surface of the car park (for example loose gravel, sand, earth are not recommended).
- Dropped kerbs to pavements adjacent to designated accessible bays.

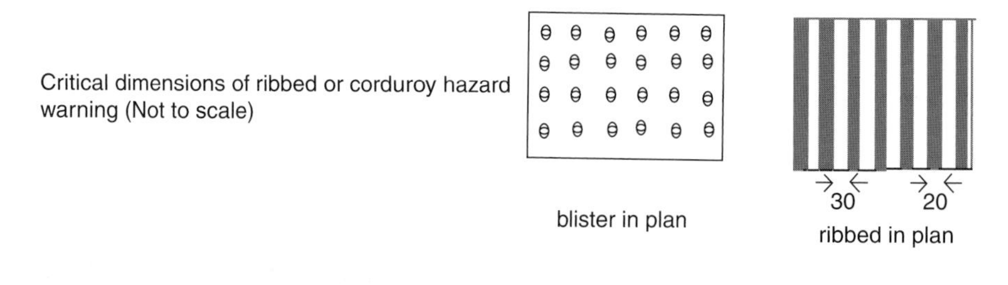

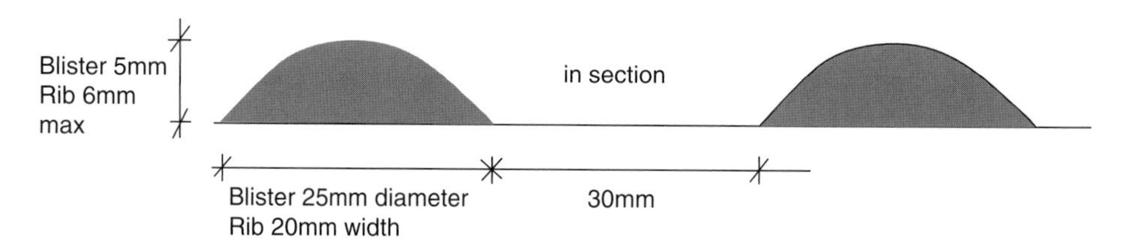

Figure 4.1 Tactile hazard warning paving

- Dropped kerbs adjacent to setting down points.
- Sufficient space at the end of designated parking bays to allow people to retrieve or replace items into the boot of a car without risk of harm from passing vehicles.
- Designated places for people with mobility impairments close to principal or alternate entrances.
- Suitable siting of ticket dispensing machines, and machines being of appropriate design for people to use who may be of short stature or in a wheelchair.
- Appropriate design of vehicle control barriers to allow people with different abilities and impairments to use.
- Appropriate numbers of designated parking bays for disabled people.
- Signage in car parks should be clear, with good contrasting colours to aid visually impaired people.

Buildings that are travelled to by car, need designated parking places for a disabled motorist. It is important to consider that these bays are clearly marked, including sufficient space to access the vehicle. Access to the building from the car park should ensure that there are no potential hazards. From leaving the car, pathways located near to the parking bay should ease access to the building. These pathways should include dropped kerbs, which aid wheelchair access and also remove a potential tripping hazard.

In ADM, it is recommended that the number of parking bays for buildings offering shopping, recreation and leisure facilities to be:

- One space for each employee who is disabled
- Plus 6% of the total capacity of the car park for visiting disabled users.

The Approved Document suggests that the provision of at least one designated accessible car parking bay on firm and level ground provided near to the principal entrance will satisfy M1 or M2 (access to a building or access to an extension of a building). It also points to BS 8300 for "the provision of parking bays designated for disabled people in different building types". This could be construed as contradictory as BS 8300 generally recommends more than one bay. The designer should therefore consider the provisions of both guidance documents, with the knowledge or expectation of the use or proposed use of the building or extension.

There are slight variations to these figures depending on the type of service the building is offering and further guidance can be sought in BS 8300, which states the following:

For general workplaces, a designated car parking space should be provided for each person who is a disabled motorist and for other disabled motorists visiting the building. When designing a building these numbers are unlikely to be known as absolutes. However, the client will not wish to have to redesign the car park once the building is in occupation and therefore the best estimate should be taken. The following are the minimum proportions of places recommended and the need may be for a higher proportion. As the obstacles in the built environment and transport systems are gradually removed allowing a greater proportion of the population access to travel with ease, so the numbers of designated car parking spaces will need to be increased.

For workplaces where the number of employees who are disabled motorists is known, the minimum number of designated spaces is recommended as one per each of these employees plus at least one space or 2% of the total capacity, whichever is the greater, for visitors.

For workplaces where the number of employees who are disabled motorists is not known, at least one space or 5% of the total parking capacity should be designated as parking for disabled motorists whichever is the greater.

For shopping, recreation and leisure facilities a higher percentage of users are likely to be disabled motorists and therefore the required number of spaces for disabled motorists should be one for each known employee who is a disabled motorist plus 6% of the total parking capacity.

For railway carparks, the minimum number of designated places for disabled motorists should be one space for each such employee plus 5% of the total parking capacity.

For places of worship, at least two designated parking spaces for disabled motorists should be provided where there is a carparking facility.

For crematoria and cemetery chapels, a minimum of two designated parking spaces for disabled motorists should be provided as close as possible to the assembly point of the crematorium or chapel.

Other places should be considered in their similarity to these examples given by BS8300. For example village halls may be similar to places of worship or leisure facilities, hotels are likely to be leisure facilities but may need a greater proportion. Provisions required at day centres and hospitals are likely to be much higher.

These figures are summarised in Table 4.1.

Table 4.1 Recommended provision of accessible parking

Building type	Number of accessible spaces	Minimum	Comment
Workplaces with known disabled motorist employees	1 per employee who is a disabled motorist + 1 space or 2% of total parking capacity whichever is greater	1 space	Differentiate spaces for employees from visitor spaces
Workplaces where the number of disabled motorist employees is unknown	1 space or 5% of total parking capacity whichever is greater	1 space	
Shopping, recreation, and leisure facilities e.g. shopping centres, theatres, cinemas, sports stadia, hotels, leisure complexes, parks	1 per employee who is a disabled motorist + 6% of total visitor parking capacity	1 space	There may need to be more spaces at hotels and at sports stadia which specialise in accommodating groups of disabled people
Railway car parks	1 per employee who is a disabled motorist + 5% of total visitor parking capacity	1 space	
Places of worship and similar religious buildings		2 spaces	
Crematoria and cemetery chapels		2 spaces	These should be as close as possible to the assembly point of a crematorium or cemetery chapel
Day centres, hospitals, clinics	1 per employee who is a disabled motorist + more than 6% of total visitor parking capacity	1 space	The function of these buildings is to provide services to a population that includes a greater than normal proportion of disabled people. Base the number on experience.

4.2.3 Accessible parking bays

A motorist may have to make a number of manoeuvres to leave their car:

(a) Park, fully open the driver door, swing legs out of car, retrieve folded wheelchair from back seat space, set it up beside the car, transfer into the wheelchair, close car door; or

(b) park, open car door, swing legs out of car, retrieve walking aid from rear or passenger seat, with the aid of this get themselves out of the car, travel to the rear of the car, open boot, retrieve wheelchair from boot, set it up, transfer into wheelchair, place walking aid in car boot, close boot; or

(c) park, get out of car and travel to rear of car, open boot, retrieve wheelchair from boot, set it up, wheel it to the passenger door, open passenger door fully, position the wheelchair, aid the passenger to transfer from the car into the wheelchair, push passenger in the wheelchair onto the footpath or other safe place, close the car door; or

(d) park, get out of car and travel to rear of car, open rear tailgate door, use rear hoist to help disabled passenger in wheelchair out of rear of car, position them safely behind the car, close tailgate.

Therefore an accessible parking bay should be of sufficient size to allow a motorist to park their car, fully open their door or the passenger door (but not both at once), and have sufficient wide safety zone at the rear of the car.

A suitable sized bay should be 4800 mm long and 2400 mm wide with a 1200 mm accessibility zone between and at the end of each bay. Like pathways, the surface should be durable, slip resistant and easy to manoeuvre on.

It is therefore recommended that in addition to the usual bay size of 4.8 m by 2.4 m a marked access zone with a width of 1.2 m is provided on one side of the standard bay and to its rear. This is shown in Figure 4.2.

4.2.4 Ticket-dispensing machines

Although not previously considered to be under the remit of building regulations, if a person needs to purchase a ticket to be permitted to park, or to be able to leave a car park for instance, if that person cannot access the machine they will have unreasonable difficulty in accessing the building. An inaccessible or unusable

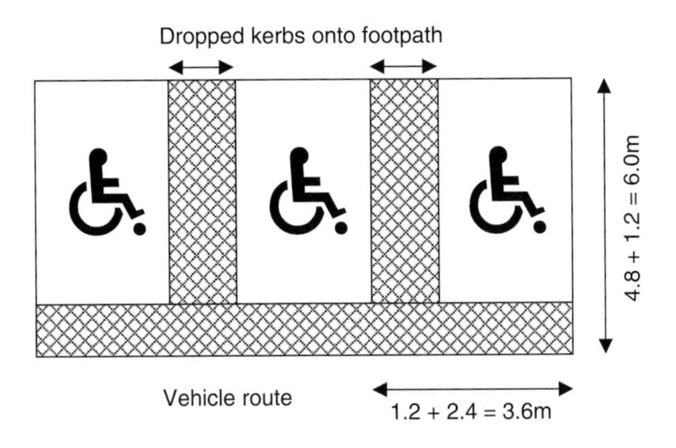

Figure 4.2 Critical dimensions for accessible car parking

ticket machine would therefore be in contravention of M1, which requires that "reasonable provision shall be made for people to gain access to and use the building and its facilities".

BS8300: 2001 is referenced in ADM as providing advice on this aspect. This is held at clause 4.1.5. Key dimensions are given in both documents so that the zone for control buttons, coin slot and ticket dispenser is positioned with its lower edge at 750 mm above ground level and does not extend above 1200 mm. The ground level area adjacent to the ticket machine should give 2100 mm by 1850 mm clear level space. This space would be used for approaching the machine, acquiring the ticket, then turning and returning to the car. Any plinth below the machine should not obstruct this area; preferably any plinth should not project beyond the face of the machine.

4.2.5 Ramped access

Despite ramps being important for wheelchair users, they are not necessarily beneficial to all disabled people. In fact, elderly people and ambulant disabled people can find ramps particularly difficult to negotiate, due to their limited movement and general unsteadiness. Even wheelchair users can find certain ramps difficult for a number of reasons:

Too steep – Making it awkward and difficult to manoeuvre up the ramp, and difficult to control while descending.

Too long – It is important that people can stop frequently in order to regain breath and rest.

No handrails – Handrails are essential for ambulant disabled and elderly people in order to use a ramp.

Through the appreciation of these problems, ramps should include the following (see Figure 4.3):

- A suitable surface that is non-slip, and is a colour that will contrast from landings for visually impaired people.
- A gradient of maximum 1 in 20, however 1 in 15 is permitted if the length of the flight does not exceed 5 m. For ramps no longer than 2 m a gradient of 1 in 12 is acceptable.
- A surface width of at least 1500 mm.
- Landings at the foot and head of the ramp of 1200 mm, clear of any obstructions, door swings etc.
- Intermediate landings that are at least 1500 mm long and clear of opening doors and other potential obstructions.
- When a user would not be able to see from the start of the ramp to the finish, landings should be a minimum of 1800 mm wide and 1800 mm long. This is necessary to act as a passing place. This size of intermediate landing is also needed when a ramp has three flights or more, because a user is more likely to tire and need to rest while others pass them.
- Unless under cover these landings should include a cross fall gradient of no more than 1 in 50 to help drain surface water.

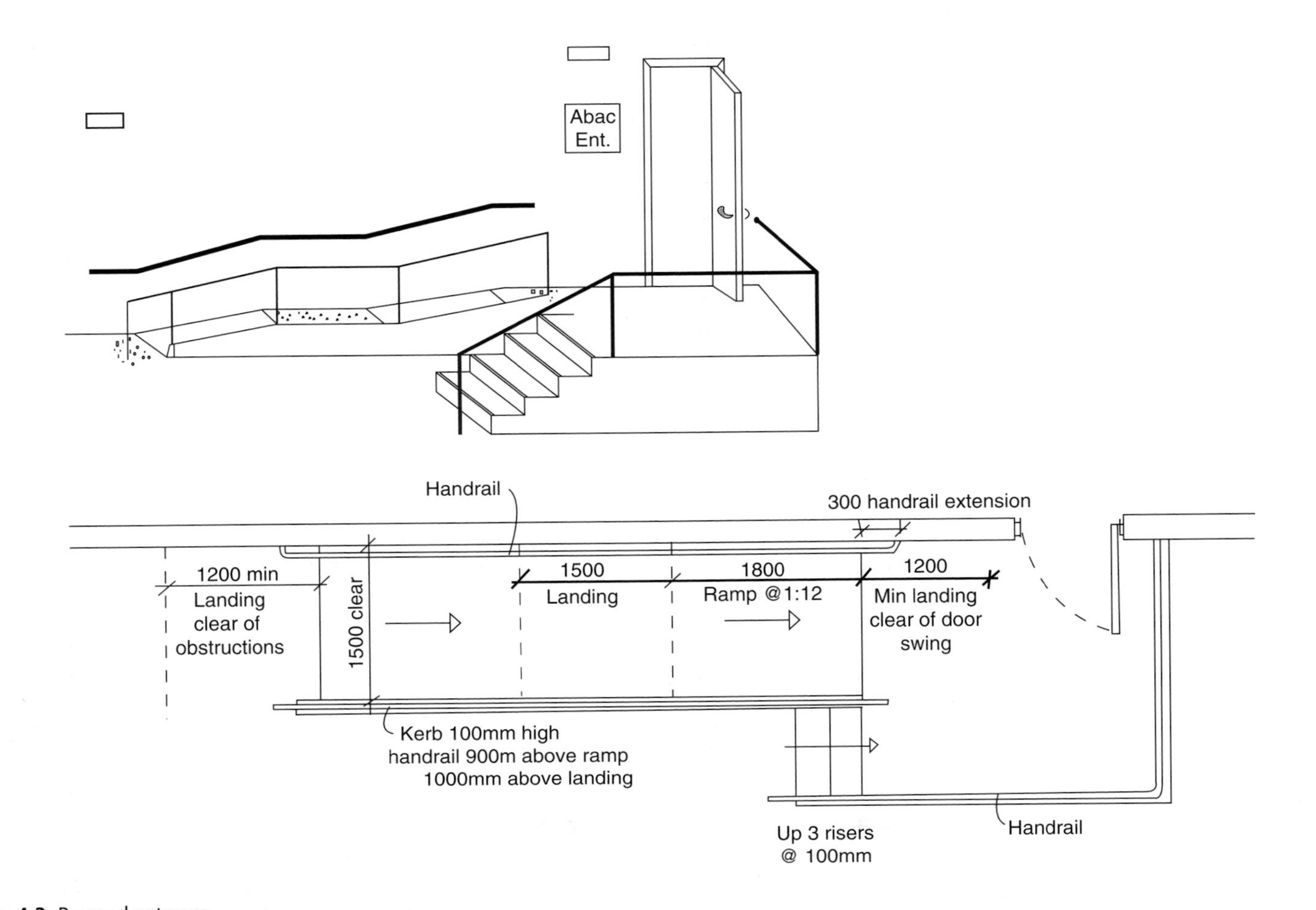

Figure 4.3 Ramped entrance

- Handrails, on both sides of the ramp, which are easy to grip.
- A kerb should be located on the open side(s) of the ramp at least 100 mm high. These kerbs should be visually contrasting from the ramp itself.
- Steps should accompany any ramp with a rise in excess of 300 mm. These should be clearly signposted.
- Adequate lighting that where possible avoids glare and excessive pools of light, which visually impaired people find difficult in terms of distinguishing changes in gradient.

4.2.6 Temporary ramps

Where buildings have limited space available for the addition of ramped access, it is possible to use a temporary ramp. In such circumstance, the ramp should be identified and marked clearly to not create a trip hazard to passers-by. The ramps should be a minimum of 800 mm wide, with a slip-resistant surface and kerbs to prevent falling off. These ramps however should not be considered in new building construction.

4.2.7 Stepped access

Although ramps are beneficial to wheelchair users, steps are particularly important, for negotiating changes in level. Suitable provision of stepped access is explained below (see Figure 4.4):

- Similar to ramped access, landings are required at the top and bottom of each flight of stairs of minimum length 1200 mm.
- These landings should include a corduroy hazard warning strip to warn visually impaired people of the potential hazard. Tactile surfaces at the top and bottom of steps should not themselves form a trip hazard.
- Steps should have a surface width of at least 1200 mm.
- Each flight of stairs must not exceed 12 steps where the going is less than 350 mm. In circumstances where the steps have a going in excess of 350 mm a flight of stairs must not exceed 18 steps.
- The rise of each step should be between 150 and 170 mm.
- The rise of steps within a particular flight should be uniform. No steps should have a rise or going larger or smaller than another. Similar to this, where there are several flights of stairs, these flights should all be consistent in terms of the number of risers.
- Steps should not overlap each other; however in circumstances where this is necessary, the overlap should be no more than 25 mm.
- Risers should not be open, i.e. without backs so that people do not trip when negotiating the steps, as toes can be easily caught.
- Handrails should be located on both sides of the steps including landings (further information on handrails is given below).
- Nosings of individual steps should contrast in colour from the rest of the tread and riser.

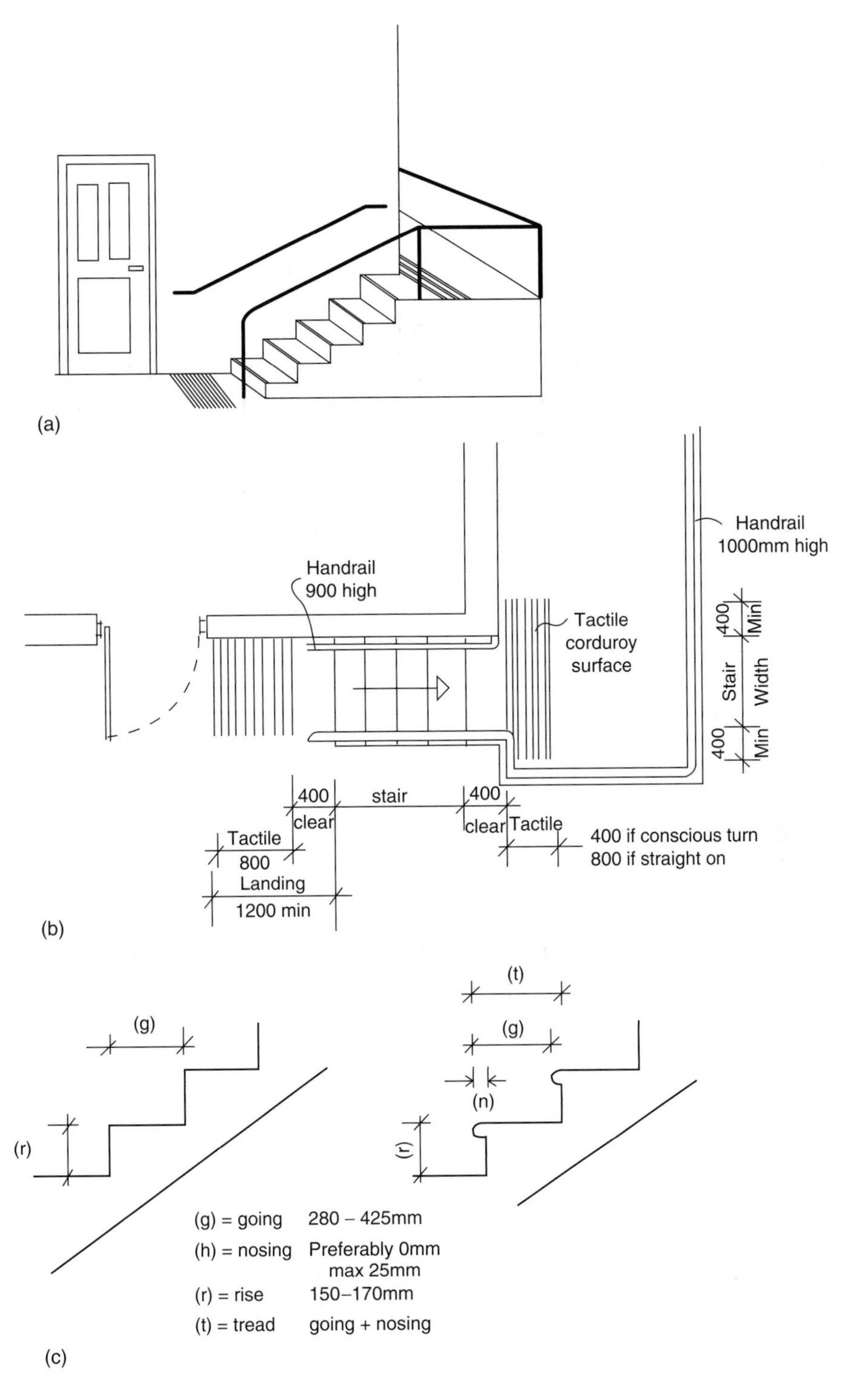

Figure 4.4 External stepped access

4.2.8 Handrails

Handrails are essential for many people in terms of negotiating a change in level. The handrail should not be placed there just as an after thought, and it should be considered an important aspect in terms of access and use of buildings. The design is necessary to suit not only people with mobility problems but also others such as elderly people and those with arthritic conditions.

The design should be suitable for people of all heights, so consideration should be given to a second lower handrail, and differing dexterity issues. People can have problems of grip, need good forearm support, or be sensitive to heat or cold. Handrails should also be spaced away from the wall surface to permit space for the hand and digits to allow the rail to be properly gripped. The supports should not get in the way of people using the rail and rails should be continuous and also extend at the top and bottom of a stair or ramp. This extension, especially at the top, is important for elderly people and arthritic people, for example, who may not have sufficient spring in their knees, ankles and feet to push up the final step and need the handrail to extend so that they can use this for leverage. The extension at the bottom is important for people with reduced balance for instance who need to hold on to the rail at the bottom of the flight to get them safely onto level ground.

Key considerations for suitable handrails differ slightly between available guidance. A summary of the differences are shown in Figure 4.5 and Table 4.2. The following presents a reasonable design guide:

- A height between 900 and 1000 mm from the pitch line on a flight of stairs or ramp, while between 900 mm and 1100 mm on landings.
- Provision of a second lower handrail where the height should be 600 mm from the pitch line of a ramp of flight of steps.
- For handrails of circular profile, a diameter between 40 and 45 mm, preferably, and no greater than 50 mm. Oval handrails should have a width of 50 mm and depth of 38 mm. The edges should be rounded with a radius of at least 15 mm.
- A clearance between the wall and the handrail of 60–75 mm.
- They should not encroach into the required width of the steps or ramp by more than 100 mm. This has implications for Part B of the Building Regulations with regard to fire safety.
- Provide a good visual contrast between the handrail and its background, with a material that is slip resistant and not cold to the touch. It should be smooth with no sharp edges.
- Handrails should extend 300 mm past the end of any landing or steps, with a design which does not permit clothing to be caught.

There are some differences between the dimensions of handrails and clear space for people's hands between the difference guidance documents (see Figure 4.5 and Table 4.2). Many advising diagrams suggest that a circular section handrail is the one to choose, however on reading the texts further, research has shown that oval or other non-circular handrails with broad horizontal faces are as easy to grip as circular rails but give better hand and forearm support.

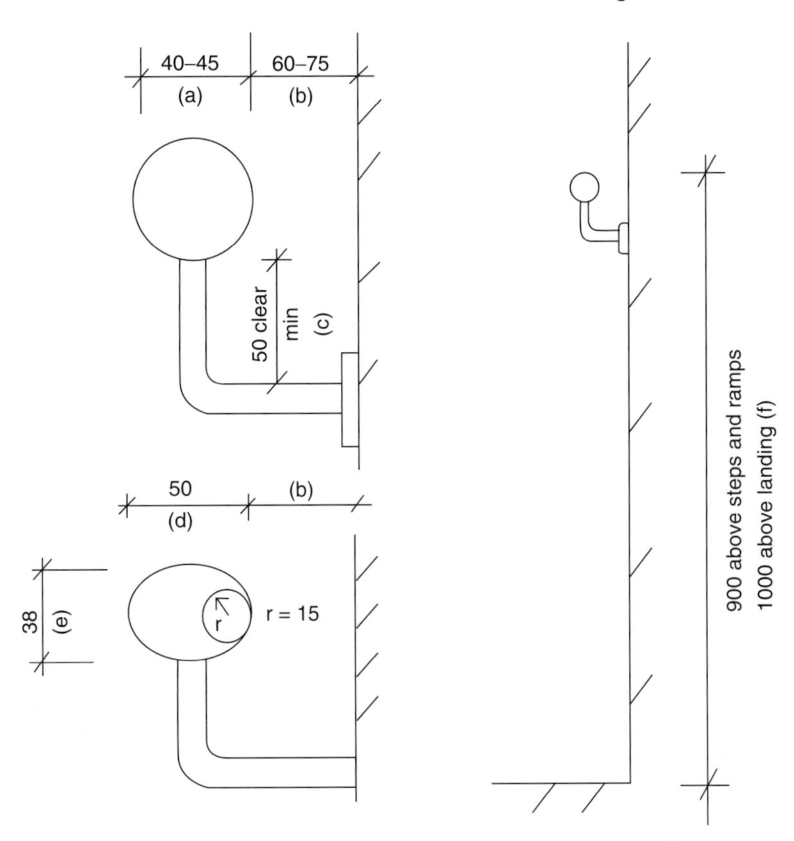

Figure 4.5 Handrail design

Table 4.2 Variations in critical handrail dimensions

Dimension (see Figure 4.5)	ADM guidance	BS8300:2001 guidance	Cae guidance (cae,1999)
a	40–45 mm	40–50 mm	45–50 mm
b	60–75 mm	50–60 mm	45 mm/min
c	50 mm/min	50 mm/min	50 mm clear
d	50 mm preferably	50 mm preferably	No guidance
e	No guidance for 'e' radius=15 mm	38 mm preferably radius=15 mm	No guidance
f	900–1000 mm above pitchline 900–1100 mm above landings	900–1000 mm above pitchline 900–1100 mm above landings	900 mm above pitchline 900–1100 mm above landings

4.2.9 Hazards on access routes

In areas whereby a circulation route passes close to a building, it is important to consider any projections, which might particularly be a concern for people with sight impairments. Common occurrences of this are windows and doors that open out onto access routes. In such circumstances, incorporating a barrier or kerb that can be detected at ground level by a cane is likely to satisfy requirement M1 and M2 of the Building Regulations.

Bollards should be carefully sited and be in contrast to any background (paving, foliage, winter foliage etc.).

Care should be taken if low walls, or fencing, including slung chain-link fencing is considered as these can constitute trip hazards.

4.2.10 Access into the building – entrance doors

Whilst designing access to the exterior of the building it is necessary to consider how a person enters the building. For all new construction the aim is to ensure that all entrances to the building are accessible without discriminating against any individual or group of people. However in the case of existing buildings this is not always possible. Owing to structural features of the building certain openings cannot be altered, yet it is important to provide an alternative accessible entrance. However this alternative must not discriminate against any person stated under the DDA 1995. For example, if the only way for a wheelchair user to gain access would be for them to have to manoeuvre to the rear of the building, then this arrangement may not be acceptable.

All main entrances should be clearly signposted, and contrast well in colour with the building envelope to aid visually impaired people. The door openings should incorporate a level-landing surface. ADM recommends a 1500 mm × 1500 mm clear area, measured from the doors being fully opened. The threshold is the crossing point from the external floor to the interior. Where possible, these thresholds should be level, as any raised surfaces can be considered a tripping hazard. However if a threshold does create a lip, this should be no more than 15 mm. Door entry systems, should be usable by all disabled people including those who are deaf, and people who cannot speak. In circumstances where there are manual doors it is recommended that weather protection is provided, such as a canopy or porch. Thought should be given to the types of surfaces at accessible entrances. Bristle mats are difficult for wheelchair users to manoeuvre on, and provide tripping hazards. Hard and slip resistant surfaces are beneficial to all users. Any mats should be fitted level with floors to reduce the chance of tripping.

Doors

Accessible entrance doors can either be manual or power operated, however for fire safety may need to be capable of being closed when not in use.

Manual Doors

In order to ensure the doors are closed when not in use, it is important that they are equipped with a self-closing device. The device should be strong enough to ensure the door will close during strong winds, but still be operable by people with weak upper body strengths. Recommendations suggest that the door should be operable by a force no greater than 20 N. A 300 mm clear space is located between the leading edge of any door and a return wall so that wheelchair users can get to the door handle and operate the door. When no space is provided, it is difficult to manoeuvre the door.

Suitable door widths vary depending on whether it is a new building or an existing building. In some circumstances, it is not possible to provide the suitable width in existing buildings, due to the building envelope and other surrounding features. Where this is an issue, reference should be made in the access statement. For certain service providers, it may not be economically viable to alter the entrance of a building. Under the DDA 1995 changes to a property have to be made only to the entrance where reasonable. If the service provider risked losing his business due to having to make large structural changes, then this maybe considered unreasonable, and alternatives to the problem can be sought.

Recommended clear widths through manual and powered-operated doors are highlighted in Table 4.3.

Where possible, when walking towards a door, the person should be able to see people walking in the opposite direction (see Figure 4.6 for recommended areas of glazing in doors). The aim is to ensure that both parties have sufficient time to react upon walking through the doors. Glazed doors can help with this, however, it is important to limit the reflection on the glass as this creates problems for the visually impaired. Similar to other accessible problems, where it is not possible to accommodate a clear view in both directions then this should be included in the Access Statement.

ADM mentions that where a latch is provided, it should be operable with one hand using a closed fist. This is so that people such as those with dexterity problems and the elderly can easily operate the door without having to form a tight grip.

Power-operated doors

Power-operated doors require many of the provisions necessary with manual doors. However, the powered movement of the door has many further implications towards gaining access. Most powered doors are activated by the use of a button or in some instances motion detectors. Doors should be set to allow

Table 4.3 Minimum clear widths of doors

Approach	New buildings (mm)	Existing building (mm)
External doors for use by general public	1000	775
Internal doors	800	750

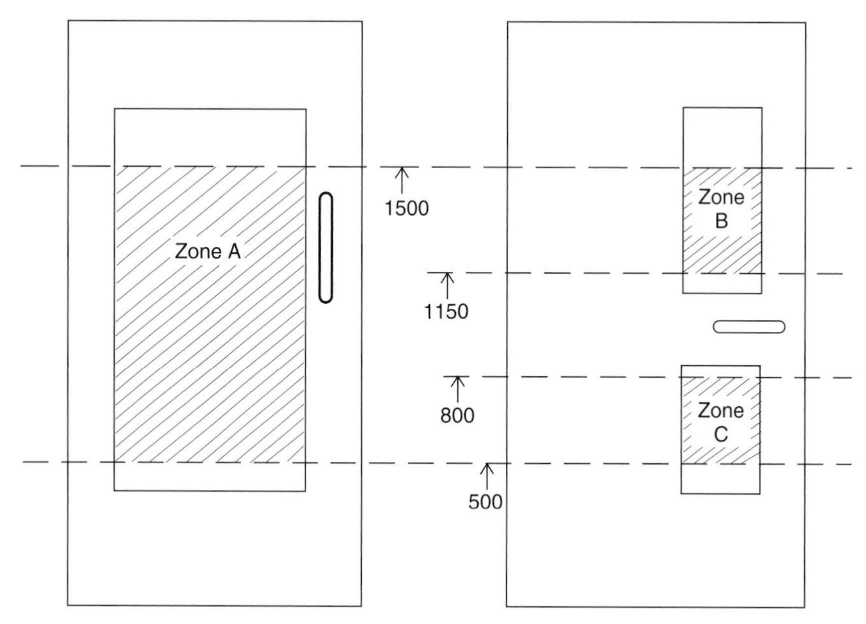

Figure 4.6 Zones of visibility for doors

adequate time for people to pass through, while sudden movements should be avoided due to people's slow reactions. Sliding, swinging or folding doors are an acceptable choice of automated door, however revolving doors are not. A normal door should always be provided alongside. This is due to the many difficulties they provide for a varied number of people with disabilities. Wheelchairs and pushchairs find it particularly awkward to negotiate revolving doors, and the risk of injury due to the continuous movement creates many concerns in terms of accessing the building. Swinging doors that open towards people should include visual and audible warnings purely as a warning of the specific automatic operation. In the event of a fire the doors should be able to be operated manually in order to allow evacuation from the building.

Control panels for the operation of automated doors should be located between 750 mm and 1000 mm from the floor level. These panels should be easy to read and be operated with a closed fist. The panel should also be placed at least 1400 mm from the leading edge of the door, to avoid any contact with the person, as the door opens.

Glazing

Glazing can cause particular problems to people with some visually impairments and it is important that glazed doors have certain provisions for its users. Problems can occur from people walking into the door and so glazed sections should have clear markings at two levels of the door between 850 to 1000 mm and 1400 to 1600 mm. (see Figure 4.7).

(a) Are these openings or are they fixed glazed panels, or is there a door? Which is the door?

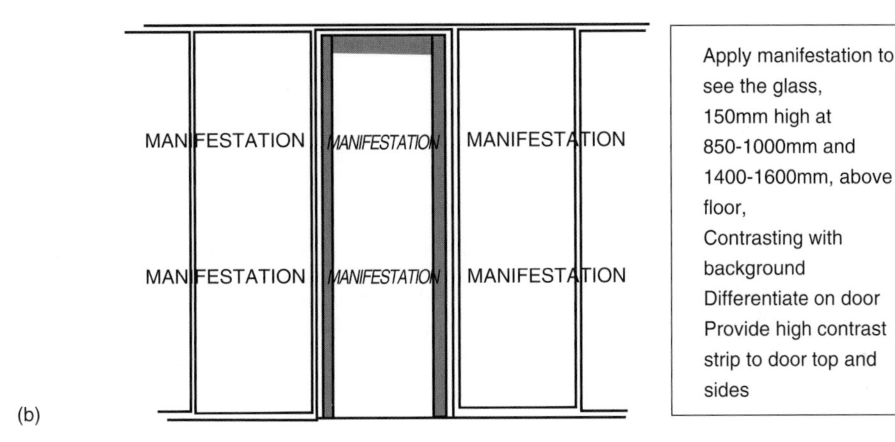

(b)

Figure 4.7 Glazing manifestations

Other problems occur where there are glazed doors in a glazed screen – finding which is the door can be difficult, especially on double swing doors with no handles.

These markings should be clear not only in terms of distinguishing from the glass, but also from the backgrounds visible through the door. The size of the markings is important to ensure that they can be clearly seen. For this reason logos should be a minimum of 150 mm high.

The door frame should also be clearly identifiable from the glazing, outlining the size and shape of the door. This is often achieved through contrasting colour, or difference in material.

4.2.11 Lobbies

Lobbies are an increasingly common feature in buildings, in order to increase security, and also reduce the heat lost through a building. When considering a lobby, attention should be made to the size and shape of the structure, allowing for suitable wheelchair access, and the provision of an additional helper (see Figure 4.8).

Where lobbies have opening doors, the minimum length (measured while doors open fully) should be 1570 mm. The minimum width for a lobby is either 1200 mm or the door width plus 300 mm.

Where there are double leaf doors, the lobby should be a minimum of 1800 mm.

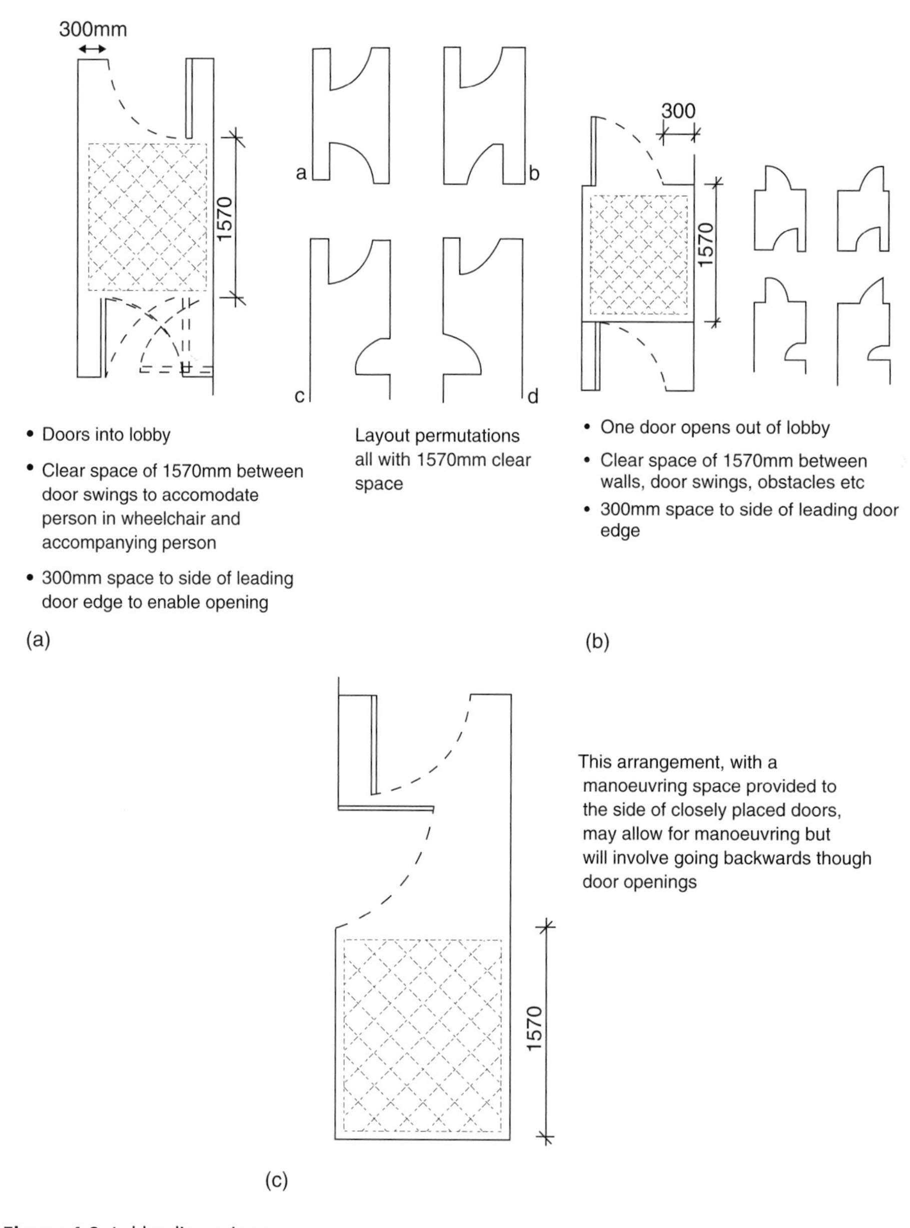

Figure 4.8 Lobby dimensions

These provisions are also required when considering any lobbies inside the building, for example, to sanitary accommodation, or lobbies off stairways.

4.3 Circulation within the building

4.3.1 Movement around the entrance storey of the building

On entering the building, for most buildings, the first point of contact is often the reception area and entrance hall. When manoeuvring around a building, there are many features that are common throughout in terms of accessibility such as:

- Clear distinguishable backgrounds between separate features to aid visually impaired people
- A solid floor surface that is slip resistant, for all users
- Easy to use controls including door handles
- Clear signage.

In addition to these key features, independent parts of buildings have further requirements set out in ADM.

4.3.2 Reception area and entrance halls

As this is the first point of contact, reception desks should be easy to get to with limited obstruction, ideally away from the main entry doors, as exterior noise can be a potential problem to those hard of hearing and also the elderly. In terms of manoeuvrability, a minimum space of 1200 mm deep by 1800 mm should be provided infront of the desk, with a knee recess of at least 500 mm deep, to allow for wheelchairs to get close. However when it is not possible to provide a knee recess, then an area of 1400 mm deep by 2200 mm wide should be provided, in front of the reception desk. The counter should include a section minimum 1500 mm wide, with a surface no higher than 760 mm for use by wheelchair users. The knee recess (if provided) should be no less than 700 mm above floor level, to allow for comfortable leg access underneath. A hearing loop should be included at any reception point for hearing-impaired people, so that they can hear and understand. This system works by allowing those with hearing aids to be able to adjust the setting on the ear piece, and communicate with the receptionist.

4.3.3 Internal doors

Similar to manual doors, when gaining access to the building there are a number of requirements that should be met. The opening force of the door at the leading edge, should be no more than 20 N. The clear width through a single leaf door is given in Table 4.3, Minimum clear widths of doors. Internal fire doors can be held open with electro-magnetic devices to allow safe access through, however should self-close upon:

- Being activated by a hand-operated device, positioned at a suitable height.
- Smoke detectors being activated.
- The result of a power failure.

When considering the opening force of doors which are fire doors, or other doors with self-closing devices, care must be taken to select an appropriate closer which will hold the door closed but not need excessive force to open it. Detailed advice is given by the Office of the Deputy Prime Minister at http://www.odpm.gov.uk/stellent/groups/odpm_buildreg/documents/page/odpm_breg_037542.hcsp

4.3.4 Corridors and passageways

Where there are corridors or passageways located within the building, it is essential that they cater to not only wheelchair users but all disabled persons. Despite the key considerations mentioned above, an appreciation should be made of the acoustics of the area to aid some visually impaired people. Sound can be used to appreciate. Soft surrounds, such as fabrics can reduce reverberation and improve the ability of a person to hear more easily. Obstacles in hallways should be avoided at all times, such as radiators and window ledges. Any projections should be clearly marked with good contrasting colours. The minimum width should be 1200 mm along a passageway. In circumstances where the passageways are less than 1800 mm wide, passing places should be included in order to allow wheelchairs to pass. These passing places should be a minimum of 1800 mm long, and ideally should also be located at junctions of corridors.

In most cases corridors tend to be level however it is not uncommon to see shallow slopes. Where a passageway is constructed with a gradient of more than 1 in 20 then this can be classed as a ramp. In such cases, reference should be made to ramp requirements. Doors opening into the corridor, when fully open should not protrude into the overall width of the passageway. This is important as otherwise this may hinder movement. Floors should be kept consistent throughout, avoiding tripping hazards, and making sure changes in level are avoided.

4.4 Vertical circulation within the building

4.4.1 Design objectives

For vertical circulation in buildings with more than one storey, lifts are an essential amenity for many people including people who use wheelchairs, people with heart conditions, and those with arthritis. The minimum standards now recommended for satisfying the Builidng Regulations is therefore the provision of a passenger lift wherever possible. The Approved Document states this in three places emphasizing the point. Therefore to provide a lift should from now on be the norm in buildings of two or more storeys. The Approved Document recommends that a passenger lift is provided in all new buildings, and in altered buildings subject to Regulation 4 of the Building Regulations, where access to a floor or level above or below ground level is being altered.

However, the common solution of installing a passenger lift will not always be possible due to building constraints and there are a number of other options available that can provide vertical movement. Notwithstanding the choice of means chosen for vertical movement, a number of key requirements are common throughout. They are:

- Adequate access and manoeuvrability space in front of the lifting device. This would be the standard 1500 × 1500 mm, or a straight access route of 900 mm wide
- Clear signage identifying the location of the device. These signs should furthermore identify the number of each floor clearly through appropriate font and visually contrasting colours
- Regardless of the type of device chosen, internal stairs should always be present as an alternative means of access, and also to be used in the event of any emergency.

4.4.2 Lifting devices

Lifting devices other than passenger lifts are an option available in circumstances when there are particular constraints on a building – such as the building being too small or maybe being classed as a listed building. If consideration is to be made for using a lifting device, other than passenger lift, then this should be discussed in the access statement, notifying the reasons for its choice. It should be remembered that a passenger lift at all times is the preferred option in terms of providing vertical access.

All lifting devices must be covered by the current Lift Regulations 1997 SI 1997/831, the Lifting Operations and Lifting Equipment Regulations 1998 SI 1998/2307, the Management of Health and Safety at Work Regulations 1999 SI 1999/3242 and the Provision and Use of Work Equipment Regulations 1998 SI/2306.

Control panels for the operation of these lifts should be at a height suitable for all persons to use easily. ADM recommends a height between 900 mm and 1100 mm, at least 500 mm from any return wall to allow wheelchair persons to get close. Controls should be clear to see and easy to use. Raised text should be provided for visually impaired people, as well as Braille; not all visually impaired people are conversant with Braille. Raised lettering is easier to read by touch than indented, particularly over time when indented letters can be filled with grease and similar substances.

In using the lifting device, it is important that an unobstructed space is provided at the point of entry. This space should be at least 1500 mm by 1500 mm so that the person can manoeuvre straight onto the device. An access route of 900 mm wide is sufficient.

A handrail should be provided on at least one side, with the top part of the rail being a fixed 900 mm from the floor. In the event of an emergency, a suitable alarm system should be incorporated.

4.4.3 Passenger lifts

All lifts should conform to the Lift Regulations 1997, SI 1997/831. BS8300 also states that they should conform to BS EN 81-1 and BS EN 81-2.

Essential requirements for passenger lifts are illustrated in Figure 4.9 and include:

- Minimum car size of 1100 mm wide and 1400 mm deep. This size of car accommodates one user of a manual or electrically powered wheelchair, and one accompanying person.

- This space is unlikely to be big enough for all wheelchair users to be able to manoeuvre or turn, and a mirror should be provided in the car so that they can see what is going on behind. The mirror should not be full length as this can be

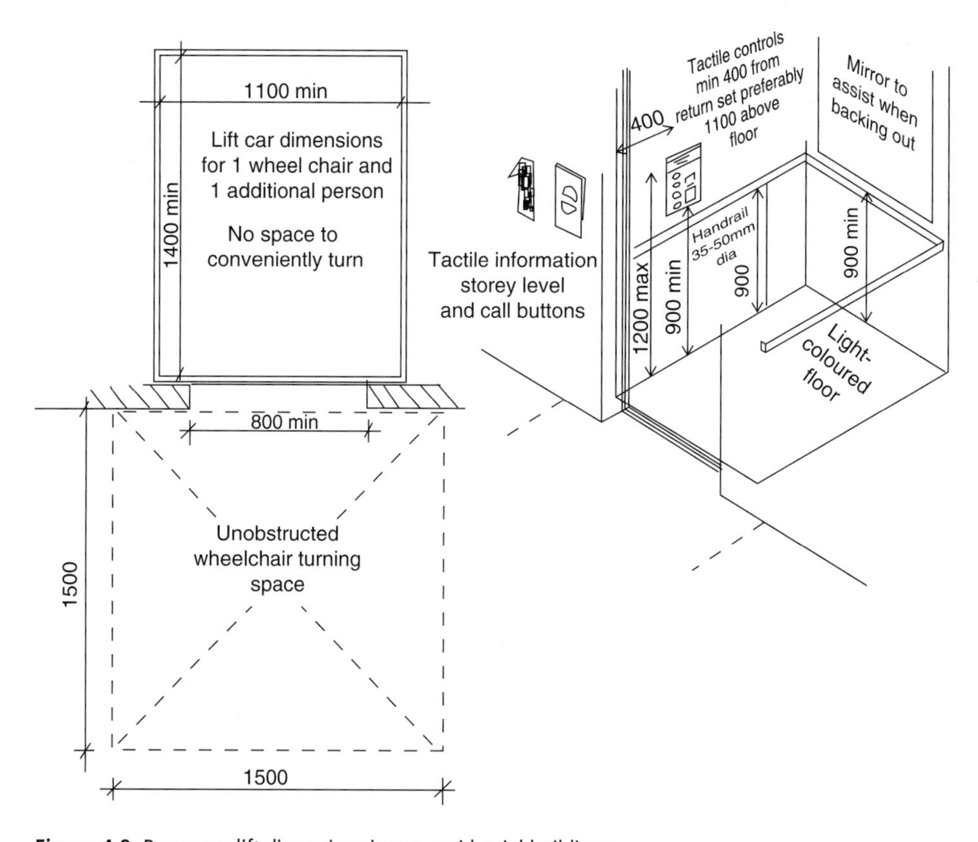

Figure 4.9 Passenger lift dimensions in non-residential buildings

confusing, as though there is another exit, and is often bronzed to avoid excessive reflection.

- The door entry should be a minimum of 800 mm wide. These doors should be power operated, and should also provide sufficient time for a person to enter and leave the cab (remember there may be disabled persons with guide dogs who require a longer period of time to negotiate the doors). Doors can be fitted with motion sensors to prevent the doors closing prematurely. The doors should also be clearly marked, so as to make them distinguishable.
- Similar to the controls on the lifting devices they should be located between 900 and 1100 mm from the ground. These controls should also be located 500 mm from any return wall.
- Audible and visual signs should be provided, indicating that the lift is at a particular floor. This system should be available at all floors and also inside the car.
- In all circumstances, a flight of stairs should also be provided in addition to the use of a lift.

BS8300 gives a second minimum dimension of 2000 mm wide by 1400 mm deep to accommodate a user of any type of wheelchair together with several other passengers. There would be sufficient room in this size car for wheelchair users and people with walking aids to turn through 180°. A car 1500 mm deep would accommodate most scooters. A car 1000 mm wide by 1250 mm deep will accommodate one user of a manual or electrical wheelchair but not with an accompanying person; this is less than the minimum guidance in ADM.

A lighter coloured floor is helpful to avoid the concern created with a dark floor that a person is stepping straight into the open lift shaft. Floors should not be reflective (see Table 4.4).

4.4.4 Lifting platforms

Lifting platforms are used mainly to transfer wheelchair users vertically from one level to another on a guarded platform. They are particularly useful for small changes in level such as between mezzanines and have been successfully used in properties as diverse as shops and cathedrals. Although designed for wheelchair users, they can be used by ambulant people also.

These systems should conform to the requirements of the Supply of Machinery (Safety) Regulations 1992, SI 1992/3073. BS8300 also states that they should conform to BS 6440.

These platforms come in many sizes in terms of the height they travel. When the system is negotiating a height no more than 2 m then an enclosure is not necessary around the lifting platform. However this is essential for safety when the platform rises in excess of 2 m. The speed of these systems is also very important and should not exceed 0.15 m/s. Lifting platform controls should follow similar guidelines to those used in passenger lifts.

Table 4.4 Internal areas of lifts

Type	Size	Accommodates	Comments
Lift	1100 wide 1400 deep	1 wheelchair + user and 1 standing person	ADM minimum car size
	2000 wide 1400 deep	1 wheelchair + user and several standing people	Given in BS 8300
	1500 deep	Most scooters	Given in BS 8300
	1000 wide 1250 deep	1 wheelchair + user only	Given in BS 8300 Does not meet ADM size requirements
Lifting platform	Non-enclosed 800 wide 1250 deep	1 wheelchair + user only	Given in ADM Only acceptable traveling less than 2m vertically and not through a floor
	Enclosed 900 wide 1400 deep	1 wheelchair + user only	Given in ADM
	Enclosed 1050 wide 1250 deep	1 wheelchair + user only	Minimum size recommended by BS 8300
	Accompanied enclosed platform 1100 wide 1400 deep	1 wheelchair + user and 1 standing person	Given in ADM

Types and dimensions for lifting platforms are as follows:

Non-enclosed platform – 800 mm wide and 1250 mm deep. Access should be a minimum of 800 m wide. This minimum internal size allows a wheelchair user only and not an accompanying person. Non-enclosed platforms should only be used where the vertical travel distance is not more than 2 m, and it does not pass through a floor

Enclosed platform – 900 mm wide and 1400 mm deep. Access should be a minimum of 800 m wide. This minimum internal size allows a wheelchair user only and not an accompanying person. If the vertical distance is over 2 m or a floor is penetrated, then the platform lift should be enclosed.

Accompanied enclosed platform – 1100 mm wide and 1400 mm deep. This type of platform lift allows for an accompanying person. Its size is also adequate when the platform has two doors at 90° to each other. Access to a platform of this size should be a minimum of 900 mm wide.

These figures differ from those given by BS 8300, which suggests minimum clear dimensions for the platform of 1050 mm wide by 1250 mm long.

All platforms should be provided with instructions of its use, which are easy to read and understand. The moving doors that provide access should be clearly visible in contrast to surrounding walls to aid visually impaired people, while an audible and visual system should be provided similar to those used in passenger lifts.

4.4.5 Wheelchair platform stairlifts

Stairlifts travel up the pitchline of a stair and can be in the form of a platform or a chair. The ADM does not consider chair stairlifts in this section. Chair stairlifts take people who have reduced mobility between floors in private dwellings and occasionally, for a specific employee, in a workplace. To use a chair stairlift, a person who uses a wheelchair has to transfer from their own chair to the chair stairlift and then to another wheelchair on the next floor. This would not be appropriate in a public building, and many wheelchair users are unable either to leave their own chair, or to use another.

Wheelchair platform stairlifts are similar to lifting platforms, except that platform stairlifts are specifically used solely for the manoeuvring of those in a wheelchair. Accompanying people and ambulantly disabled people cannot use a platform stairlift. The person travels up and down seated in their own wheelchair, which is rolled on and fixed to the device during travel. These are not a preferred option and a passenger lift should be favoured over the installation of such a system.

The system should comply with the requirements of the Supply of Machinery (Safety) Regulations 1992, SI 1992/3073. These lifts are often used infrequently and it is essential that on a single flight of stairs that the lift when in the parked position does not encroach onto the stairs minimum width. In order to keep the disabled user safe the stairlift should not exceed a speed of 0.15 m/s. Controls for the lift should be similar to those used on a passenger lift.

Dimensions should be a minimum of 800 mm wide by 1250 mm deep. The minimum width for access should be at least 800 mm wide.

Care is needed to ensure that there is no conflict between the stairlift, either in use or folded away, and the means of escape on the stairs particularly with regard to the remaining clear width of the stair and the loss of the handrail. If the stairlift is being used to travel along the course of a flight of stairs, the lift should not obstruct the use of stairs by other people. This would necessitate a much wider stair.

4.4.6 Internal stairs

Internal stairs follow similar guidelines to "Stepped access" in that many of the requirements are the same. However ADM does include a number of further measures that should be looked at when considering internal stairs.

Each individual flight of stairs should not exceed 12 risers, without the inclusion of a landing in order to allow a resting area. The risers must also be between 150 and 170 mm, so that it is easy, particularly for those with movement difficulties, to manoeuvre up and down the stairs. The going of the steps is recommended to be not less than 250 mm. This is to ensure that there is plenty of room for placing a foot and negotiating the flight.

In previous editions of the Approved Document, a hazard warning surface (tactile surface) was required at the head of internal stairs. While these are still recommended for external stairs, it is now considered not reasonable to require them internally because there is no recognized warning surface for use internally, which will definitely not constitute a trip hazard, especially when used in conjunction with flooring surfaces, which have different frictional resistance characteristics.

There is a need for research into safe internal use of tactile surfaces particularly where the top of a stair is in a direct line of travel with no other prior warning.

Although not included in ADM, the wording is that "it is not reasonable to require" tactile hazard warning to internal stairs, rather than stating that such method of warning is to be actively discouraged. The British Standard recommends that tactile warning is provided. This has resulted in confusion particularly where considering the DDA. There have been occasions where tactile warning has been provided in consideration of DDA and trip accidents have occurred. This is obviously not sensible either. There is no requirement under the DDA to provide tactile warning, as there is no requirement under the DDA to provide any prescriptive measure. For each individual situation where there is a defined need, a reasonable solution should be sought, which does not jeopardize the health and safety of any expected user. Due regard for safety should be taken and careful consideration be given to surrounding floor surfaces, including slip and friction characteristics, contrasts of colour and luminance and good lighting.

Audible warnings could be considered dependent on the use of the building but this again is not a universal panacea because the audible warning would need to be recognizable as such by the person using the stairs and in constant working condition. It is not unknown for people to be momentarily alarmed by an unexpected noise – which may also result in accident at the top of the stairs.

So, to clarify, no hazard warning surface is currently required at the top of internal stairs, but designers should be aware of the potential dangers of having a stair directly in line with an access route, and design out the hazard accordingly.

To clarify the requirements under ADM for stairs and steps inside the building see Table 4.5.

Exceptions can be made to the above when there are building constraints and such alterations are not possible. In such cases it is important to note this in the Access Statement.

4.4.7 Internal ramps

Ramps are not necessarily a perfect design solution. Especially internally, they can be a trip hazard and are not always safe or convenient for ambulantly disabled people. Their inclusion in designs should therefore be carefully considered.

Table 4.5 Critical internal steps design

Item	Dimension/ number	Comment/requirement
Number of risers in a flight	12 2 16	Maximum Minimum No single steps, provide ramp Maximum in small premises where space is restricted
Landing positions required at:		Top of flight Bottom of flight So that no flight has more than 12 risers
Length of landing (going)	1200 mm	Unobstructed No doors must swing across a landing Same width as steps
Tactile hazard warning	—	Not required to internal steps
Width of steps	1200 mm	Minimum Between enclosing walls, strings or upstands
Nosings, width on tread and riser to be:	55 mm	Nosings to be made apparent by permanently contrasting material 55 mm wide on both the tread and the riser
Nosings, projection over lower step	25 mm	Maximum Preferably avoid projections
Risers	150 mm 170 mm	Minimum Maximum In existing buildings, the case for a different rise must be argued in the Access Statement. In school buildings the rise should not exceed 170 mm, whether existing or not No open risers Risers to be consistent throughout a flight, Where total rise is less than 300 mm, provide ramp (no single steps)
Goings	250 mm 280 mm	Minimum Minimum in school buildings 300 mm minimum is preferred by people with mobility impairments Goings to be consistent throughout a flight
Handrail		To every flight and landing On both sides Continuous

(Continued)

Table 4.5 (*Continued*)

Item	Dimension/ number	Comment/requirement
Handrail height above landing	900 mm 1100 mm	Minimum Maximum
Handrail height above pitchline	900 mm 1000 mm	Minimum Maximum
Handrail length	300 mm	Minimum extension horizontally beyond the top and bottom nosing Not to project into an access route Wall/partition designs must take this into consideration
Handrail design		See Figure 4.5 and Table 4.2

Similar to internal stairs and its association with stepped access, internal ramps follow a very similar theme to external "Ramped Access" at 4.2.5. Exceptions to this however occur when the ramp rises by more than 300 mm or more. In the event of this, two or more sign-posted steps should be provided. Alternatively where the change of level is less than 300 mm there is no necessity to provide a single step.

As required with ramped access, it may be necessary to include a landing, to provide a resting point. These landings should, where possible, be flat, however, if there is a slope, this should not exceed 1 in 60.

Handrails provided for internal ramps follow the same instructions to external handrails discussed at "Approaching the Building" and in Figure 4.5 and Table 4.2.

The requirements for internal ramps are summarized in Table 4.6.

4.4.8 Areas beneath stairs or ramps

Areas below stairs or ramps where the soffit is less than 2.1 m can be a hazard, to people not paying attention and to people with visual disorders. To avoid collisions, the areas are to be guarded. This can be achieved by guarding, which is detectable by a person using a cane, or by a permanent barrier. Fixed rails can be used but must have, for example, balusters or supports at sufficient centres to be detectable by someone using a cane. Internal planting features have been used to good effect in such positions, but must not cause a trip hazard.

4.5 Aids to communication

4.5.1 Design objectives

For the purposes of orientation in a building, use of the building and its facilities, and for safety, a variety of message methods are utilized in the building. The

Table 4.6 Critical internal ramp design

Item	Gradient/ dimension/ number	Comment/requirement
Clarity		Ramp to be readily apparent or sign-posted
Gradient (maximum) of a flight in relation to its going and total rise	1:20 1:15 1:12	Maximum rise 500 mm, maximum going 10 m Maximum rise 333 mm, maximum going 5 m Maximum rise 166 mm, maximum going 2 m Or use equation Gradient = 1:x, where $x = (10 + \text{going})$
Maximum total rise	500 mm	Between flights + landings
Accompanying steps		Where rise is 300 mm or more, 2 or more clearly signposted steps Where rise is less than 300 mm, no accompanying steps.
Maximum going	10 m	Between flights. Provide a landing.
Width (minimum)	1500 mm	Surface width between walls, upstands or kerbs
Slip-resistant surface		Especially when wet
Colour		Contrasts with colour of landings (or level routes)
Frictional characteristics		Those of the ramps and the landings surface are similar
Landing positions required at:		Top of flight Bottom of flight So that maximum rise and going is not exceeded (intermediate landings)
Length of landing (going) (minimum)	1500 mm 1200 mm	Intermediate landings At head and foot of ramp Unobstructed. No doors must swing across a landing
Intermediate landings	1.8 × 1.8 m	Minimum width and going when three or more ramp flights, or cannot see from one end of ramp to the other. Permits wheelchair passing places
Landings gradient		Level Maximum 1:60 along length of landing
Handrails		Both sides
Kerb on open sides	100 mm	High to any ramp or landing Contrasts visually with ramp or landing This in an addition to guarding under Part K

building is not simply a physical presence but a communication device. To ensure communication is effective it has to reach and be understood by its intended audience. If some of this audience miss or misunderstand these messages, then communication has failed. Systems provided in buildings such as lighting, sound systems, fire alarm systems and signage can positively enhance the effectiveness of communication and utilisation of the building. The ADM considers that aids to communication will satisfy Requirement M1 if:

- a clearly audible public address system is supplemented by visual information;
- provision for a hearing enhancement system is installed in rooms and spaces designed for meetings, lectures, performances, spectator sports, films and so on and also at service or reception counters in a noisy area or they are behind a glazed screen;
- the presence of an induction loop or infrared hearing enhancement system is indicated by the standard symbol;
- telephones suitable for hearing aid users are clearly indicated by the standard ear and 'T' symbol and incorporate an inductive coupler and volume control;
- text telephones for deaf and hard of hearing people are clearly indicated by the standard symbol;
- artificial lighting is designed to be compatible with other electronic and radio frequency installations; and
- auxiliary portable devices that incorporate vibrating features can be offered or supplied as standard (e.g. pillow devices for hotel bedrooms).

These aids are mainly useful for people with hearing impairments. It can be seen that communication of the presence of facilities for disabled people is given as much importance as the provision itself. The illustration in Figure 4.10, taken from BS8300, shows standard public information symbols.

The designer will be aware that many different people can have difficulties in orientation and use of buildings and services, including non-English speaking people as well as people with sensory impairments and cognitive problems.

The way a building communicates to a person works on many different levels and the ways a person picks up messages given out by a building are also diverse and numerous. The specific requirements of ADM and other issues are discussed below. People pick up the messages by sight, hearing, touch, and resonance, and these are not mutually exclusive. Recent research on the brain shows that the brain is not organized into separate visual bits, audio bits, touch bits, spatial awareness bits etc., but that these areas, traditionally understood to be central to these individual senses, are connected in interesting ways, more like the internet than a PC. The brain adapts with every thought and action made, and if some sense is removed, other areas may be able to compensate and grow in response. For example, the visual cortex of the brain can show increasing activity in blinded people performing tasks with their fingers or when hearing tones or words. They were in fact "seeing" what they were touching or hearing, albeit not in the accepted way that sighted people see. There is a suggestion that connections from

Standard symbol for a hearing enhancement system such as induction loop which utilises the 'T' switch on hearing aids

Standard symbol for a hearing enhancement system such as infrared

Figure 4.10 Standard public information symbols

touch or audio senses to the visual cortex may exist in everyone but remain unused whilst the eyes are working. When the eyes are not working, the next best way of getting the same information is put into play. So it could be that the visual cortex is not devoted to sight but to spatial awareness.

Therefore the consequences of a designer considering how the design will aid people with different sensory perceptions are crucial in enabling a person to negotiate around a building. Consistent messages in terms of textured information, audible information, sound reflective qualities, colour tones and contrast information could all be used to great effect. It is to be hoped that further research will enhance understanding and therefore result in improved advice for designers in these areas.

In the meantime, designers should do the best they can.

4.5.2 Public address systems

If a public address system is provided, this should be clearly audible. There is little further guidance given in ADM. The BS8300 states that address systems for performances and announcements should be amplified in a form that is suitable for people with impaired hearing and where possible should be tested by user trials.

Both documents also require supplementation by visual information. This could be by light-emitting diodes (LED) displays such as those seen at football and cricket stadia, or by television text systems.

4.5.3 *Fire alarm systems*

Fire alarm systems generally have to be as required by Part B and Approved Document B, however there are some additional requirements discussed in ADM to ensure that all people are able to benefit from the earliest warning of a fire or other emergency, their safety could be jeopardized without these provisions. To incorporate these, requires an interpolation of the requirements in Part M and Part B, together with the guidance in the British Standards mentioned below, and a view to the DDA for the service provider of the occupied building; the term "joined-up thinking" could be applied here.

The current guidance for sounders and beacons in the UK is contained in BS 5839-1:2002 Fire detection and alarm systems for buildings – Code of practice for system design, installation, commissioning and maintenance. There is no UK standardized alarm tone information, or guidance on the light output of beacons. Sound output should be at least 5 dB(A) above the normal ambient sound levels, generally not less than 65 dB(A), and at 75 dB(A) at the bedhead of a sleeping person.

The lack of standardized alarm information tone, and the plethora of alarm noises people are subjected to, means that the general public cannot always appreciate whether an alarm means "fire" or something else to be ignored. In the last 20 years, particularly following the research after the King's Cross fire, voice alarms have become more frequently used. Here the alarm tones are supplemented by recorded, or sometimes active, speech messages, such as "Please evacuate the building by the nearest available exit" or even "There is a possible fire on the seventh floor, west wing. Please evacuate the building immediately by the nearest available exit, avoiding the seventh floor, west wing." Such a system allows specific instructions to be given to allow an effective evacuation. They do not have to be dedicated to fire evacuation but can be used for public address announcements or broadcasting music when there is not an emergency situation.

Speech usually results in people moving for evacuation very much faster than a tonal alarm, for example, in one test within 2 minutes rather than above 10 minutes, but it would not work in all situations for all people. To work at all, sounders must be audible and intelligible, and the message must be clear.

It must be remembered that people can have multiple disabilities or impairments and that, for example, a wheelchair user could be deaf, or an ambulant disabled person be blind. Such people may need more time to be able to make their safe escape and therefore particular care is needed to afford them the ability to appreciate that the alarm has been activated, especially in places where there may be no other people around and therefore they may not be alerted additionally by their colleagues or the movement and reactions of other people.

ADM therefore requires that in sleeping accommodation, for example, in hotels, motels and student accommodation, all bedrooms have a visual fire alarm signal, in addition to the requirements of Part B.

In sanitary accommodation in buildings other than dwellings, any fire alarm system should emit a visual and audible signal to warn occupants with hearing or visual impairments. This should be given careful consideration to determine how hearing impaired people who may be in enclosed cubicles will be alerted.

In addition to fixed systems, BS8300 suggests that vibrating devices can be used. These can take the form of wearable paging units, pillow vibrating units or under mattress pads designed to wake a person from sleep. However the unit has to be in contact with the person to be of use. It obviously does not work if left in a jacket pocket, or the person is getting dressed or showered.

Great care must be used if flashing or strobe light systems are proposed. These can cause confusion, disorientation and, in some people, epileptic fits.

Xenon beacons have always been disliked by system designers, due to their high power consumption and surge currents. These cause problems with interface systems especially addressable ones trying to cope with a number of simultaneously operating beacons.

The LED beacons are becoming more commonly used as an alternative as they can match performance at a tenth of the current used by Xenon beacons. However the LED beacons currently available are unlikely to be bright enough to wake someone from sleep or alert the attention of someone working intensely.

There is no single solution; current technology suggests an appropriate mix of voice alarm, light beacon and vibrating devices.

4.5.4 Surface finishes

Orientation in a building is obviously essential for using the facilities. All users need to be able to find their way around, including visually impaired people. The ADM explains that the appropriate choice of materials for ceilings, walls and floors can help visually impaired people appreciate the boundaries of rooms or areas. They can also help in identifying access routes and receiving information. BS8300 adds that people with visual impairments can encounter difficulties in finding their way around spaces if they cannot perceive and therefore respond to visual clues or if they find it difficult to distinguish sounds in an acoustically reverberant environment. People who use wheelchairs or walking aids or people with ambulant impairments can be unnecessarily restricted by the choice of a high resistance floor surface such as a deep pile carpet or by highly slippery surfaces.

The following characteristics, given in BS8300, affect the ability of different people to find their bearings and way around a building:

- The colour, luminance and texture of a surface
- The contrast between elements such as walls and doors, architraves, skirtings, cornices, handrails etc., which define particular areas or boundaries
- The appropriate use of surfaces to clarify location and direction and to identify objects
- The acoustic environment
- The grip of floor surfaces, particularly at changes in level.

The ceiling is usually the most uncluttered element in a room and if well contrasted with the walls or bounded by a contrasting cornice, visually impaired people can often determine the size of the room from the ceiling. However as most people concentrate their vision below 1200 mm from the floor, contrast between the wall and the floor is critical in wayfinding.

The use of surfaces that contrast with each other is usually the most effective and appropriate way of providing ease of orientation. BS 8300 explains this in greater depth, using information from the RNIB. Luminance contrast, which is a contrast of brightness, is more important than colour contrast in helping visually impaired people distinguish between different surfaces. Colour difference can help if the colours at adjoining surfaces are chosen either for the different amount of light the colours reflect or for the different intensity of the colours chosen. The use of colour from different parts of the spectrum (colour of a different hue) is less suitable than combinations chosen for both colour and luminance contrast because there are people who are insensitive to differences in hue.

Colour and luminance contrast should be used to distinguish the boundaries of floors, walls, doors and ceilings. The colour and luminance of the walls should be noticeably different to that of the floors and ceilings. High gloss and mirrored surfaces should be avoided for floor, wall, door and ceiling surfaces particularly in circulation areas as these can cause glare and/or confusion.

Glare and reflection from shiny surfaces can cause pain or confusion. Large repeating patterns can cause similar adverse visual effects. These should be avoided in spaces such as circulation areas and where receipt of visual information is important such as at reception areas and by speaker's rostrums in lecture halls.

Ceiling, wall and floors materials should contribute to an acoustic environment that helps orientation and enables audible information to be heard. BS 8233 gives design recommendations so that the appropriate acoustic absorbency can be chosen for each material. Hard materials such as wood, stone, plaster can reflect sound and create a noisy environment. High absorbency materials can result in environments that give a deadened atmosphere.

Floor surfaces with high gloss finishes should be avoided due to problems with glare and the impression of slipperiness induced. Bold patterns and patterns, which simulate steps should be avoided. Surfaces should be slip resistant under wet and dry conditions. BS 5395-1:2000 and BS 8300 Annex C give guidance on slip resistance for stair and floor surfaces and should be consulted particularly when designing for public areas. Some of the information is included in Table 4.7 for illustration.

Other floor surfaces cause problems of restriction for wheelchair users, for example, deep pile carpets. Similarly coir matting should not be used on floors or within a mat-well.

4.5.5 Hearing enhancement systems

People with hearing impairments often have difficulty hearing in larger venues because of the blurring effects of room acoustics. Other noises, such as made by

Table 4.7 Slip potential characteristics. Taken from part of BS 8300 Table C.1 in Annex C which replicates BS 5395-1:2000, Table 4. Illustration and example only. Refer to BS 8300 for more information.

Material	Potential for slip		Remarks
	Dry and unpolished	**Wet**	
Carpet	Extremely low	Low	Loose or worn carpet can present a trip hazard
Ceramic tiles (glazed and highly polished)	Low	High	If open treads are used , the potential for slip can be low in wet weather
Clay tiles	Low	Moderate to low	When surface is wet and polished, the potential for slip can be very high
Cork tiles	Extremely low	Low	–
Granolithic	Low	Moderate to low	Slip resistant inserts are necessary whenever granolithic is used for stair treads. Polished granolithic should not be used for stair treads
Linoleum	Low	Moderate to low	Edges of sheet liable to cause tripping if not firmly fixed to base
Resin, smooth self-leveling	Extremely low	High to moderate	–
Resin, enhanced slip resistance	Extremely low	Low	The anti-slip properties depend upon sufficient, uniformly distributed aggregate. Areas of reduced aggregate can present serious slip hazard
Rubber (sheets or tiles)	Extremely low	High	Not suitable near entrance doors or other foreseeably wet areas, for example, shower rooms, sanitary accommodation or kitchens
Terrazzo	Low	High to moderate	Slip resistant inserts are necessary whenever terrazzo is used for stair treads. Polished terrazzo (including resin-based) should not be used for stair treads
Timber (finished)	Extremely low	High	Applies to sealed or varnished or polished timber
Timber (unfinished)	Low	Moderate	–

people in the room, traffic outside, works going on nearby can adversely affect the receipt of the required sound. For people with hearing difficulties, these distractions can have the effect of being magnified so much that the required sound is obliterated. The problem is also made worse by the distance the sound has to travel before reaching the audience. Hearing enhancement systems are designed to enhance the sound either through the hearer's own hearing aid or by the use of separate headsets.

The ADM recommends that hearing enhancement systems should be installed in rooms and spaces used for meetings, lectures, performances, spectator sports, films and so on and also at service and reception desks or counters where the background noise is high or there is a glazed screen between people communicating.

Where these systems are installed, ADM also recommends that users of the building are informed of this provision, by use of the standard symbol in an appropriate place.

The Royal National Institute for Deaf people (RNID) produce fact sheets as part of their special equipment range (RNID, 2002). These give information for people managing public venues and further information can be found in the sheets. Much of the following is taken from these sheets and from BS 8300.

Audio frequency induction loops and infrared systems are two systems designed to overcome these problems. Systems that use induction loops and infrared replace the air sound path between the sound source and the listener with either an inductive signal or an infrared signal. Radio systems are a third option.

The systems do not improve the sound receipt for everyone in the room. The listener uses a receiver that converts the signal back to sound. Loop systems use the listener's own hearing aid, while infrared systems use special receivers, usually lent to the person for the specific event, which convert the signal back to sound.

They can be useful in a number of different situations and buildings for example theatres, cinemas, places of worship, meeting rooms, lecture halls, conference rooms, airports, railway stations, and shopping centres. As well as for large areas, counter loops can also be installed, useful for banks, ticket offices, post offices and so on, where there is a glass screen between staff and customers.

In quiet environments, where the speaker and listener are close to each other and there is no glass screen between them, an induction loop or infrared system is unlikely to be required.

Induction loops

An induction loop is a cable, which encircles the audience area and is fed by a loop amplifier. The amplifier receives the signal from a microphone placed in front of the person speaking or directly from the sound system etc. such as a television or public broadcast system. The resulting electric current in the loop produces a magnetic field corresponding to the sound of the speaker's voice. This magnetic field can then be picked up by anyone within the loop area by switching their hearing device to the 'T' setting.

In a room, the loop usually runs around the edges of the room so that it serves the entire audience area. It could, however, only circle a particular area such as a small seating area or a desk countertop. Portable loops are also available, covering a small area, can be packed away after use and are useful if a permanent system is not required or desired.

Although loop systems are not affected by background noises, hearing aid users may hear magnetic interference from electrical equipment, fluorescent lights, light dimming systems or power cables. This interference is picked up directly by the user's hearing aid.

Induction loops are also subject to over spill. Walls, ceilings and floors do not form a barrier to magnetic fields. Therefore hearing aid users outside the room or area may be able to overhear conversations or speeches from the looped area. This can be a particular consideration when installing two adjacent meeting or lecture rooms with loop systems or when confidentiality is necessary.

As electrical equipment can interfere with the workings of the loop system, so loop systems can cause interference in other parts of a sound system.

As the system is essentially electromagnetic, metal in buildings can have an unpredictable effect on loop systems. Where there is a large amount of metal, for example in a steel-framed building, a weaker inductive signal may be experienced or may vary from seat to seat.

Once known about or predicted, these difficulties can usually be overcome by specialists.

Loop systems in public buildings should conform to BS EN 60118-4, and be installed, set up and calibrated to BS 7594, the Code of practice for audio frequency induction loop systems.

Infrared systems

Infrared systems use invisible infrared light to carry sound to receivers worn by listeners. The complete system consists of infrared radiators, a pre-amplifier or mixer unit and a microphone or other audio input source. Sound is fed into the preamplifier or mixer, where it is processed and passed the radiator for transmission as infrared radiation. Radiators cast infrared radiation over the listening area.

Users need a receiver and can be positioned anywhere in the area covered by the radiators. Many receivers can be used without hearing aids and are worn as head phones, earpieces or around the neck.

Walls and other surfaces in the coverage area reflect the infrared radiation. This can mean that listeners may be able to receive the sound even if they are not directly facing an infrared radiator. However some wall coverings do absorb the infrared radiation.

Infrared systems are not usually prone to interference unless the receivers are in direct sunlight. Because the radiation cannot pass through walls, ceilings or floors, systems in adjacent rooms be used at the same time and the confidentiality consideration does not arise.

Infrared systems should not be designed to work at frequencies low enough to be affected by other infrared sources, such as high frequency fluorescent tubes.

Radio systems

Radio receiver hearing enhancement systems can also be used. These can be completely portable and are commonly used in educational establishments and art galleries where a speaker, wearing a transmitter, is moving from place to place and requires their listeners, who wear receivers, to be able to hear. The radio signals can usually be received up to a distance of 60 m.

More permanent fixtures, such as in large meeting rooms, involve fitting microphones, which provide output to radio sound aid transmitters, and sound amplifying or sound reinforcement loudspeakers.

Radio waves can travel through walls, floors and ceilings and therefore this is not a system to be used when confidentiality is an issue.

Radio systems, which allow transmitters and receivers to be switched between different channels, can be used in adjacent rooms without picking up overspill sound. However they can be susceptible to electromagnetic interference and radio signals from other sources on the same wavelength.

4.5.6 Telephone systems including payphones and entry phones, etc.

The Approved Document does not require that a telephone system is installed, nor that if one is installed it must be suitable for hearing aid users. However if a telephone system suitable for hearing aid users is installed in the building, this should be indicated by the standard ear and 'T' symbol and incorporate an inductive coupler and volume control. The inductive coupler is for people who wear a hearing aid that has an inductive pick-up (the 'T' switch) and the volume control is to adjust amplification for people who do not wear a hearing aid but have significant hearing loss.

BS 8300 recommends that these should be fitted into the circuitry of all public or visitor payphones, entry phones and emergency telephones in lifts.

Where these facilities are provided the sign shown in Figure 4.10 should be used to indicate its presence.

Similarly for text telephones in that the ADM does not require that they are provided, but where they are they should be identified with the standard symbol.

It should be noted here that where the DDA 1995 has effect, if a telephone system is supplied, this facility should not discriminate against disabled people. So facilities should be provided for people with hearing impairments as well as for people without hearing impairments.

4.5.7 Artificial lighting

Artificial lighting should be carefully designed. Good artificial lighting is essential to the health and well-being of many people in buildings. Poor lighting at workplaces can have a severe effect on employees. People with visual impairment need well-designed lighting to be able to use a building and its facilities. Similarly, people with impaired hearing need to be able to see the face of people speaking to see and understand the movement of lips for lip reading and hands when signing.

Artificial lighting systems should be designed to be compatible with other electronic and radio frequency installations.

Lighting needs to give good colour rendering of all surfaces, without creating glare, pools of bright light, or dark shadows. The colour rendering of surfaces can be improved if guidance in the CIBSE Code for interior lighting is followed (CIBSE, 1994). Uplighters positioned at or near the floor level can be disorienting, by creating glare and obscuring vision, and should be avoided.

Lighting levels can also be obtained from the CIBSE code. This states that most published recommendations for lighting levels inside buildings are on a simplified basis of 'illuminance on the working plane'. It is important to note that where task illumination, as measured on the horizontal working plane provides an important minimum specification, this is a minimum and other factors such as limiting glare index, avoidance of harshness, enough light on walls and ceilings and so on must also be met. Generally, the more detailed and exacting the work carried out in the space, the higher the level of illuminance on the working plane needs to be.

Table 4.8, showing figures and explanations taken from the CIBSE guide, gives a rough guide. These will all need to be modified dependent on the conditions.

4.5.8 Signage

Effective signage is essential for many aspects of a building's facilities. Detailed information is not given in ADM which directs the reader to BS 8300. People need clear information about the purpose and layout of spaces for orientation and of available facilities.

Table 4.8 Standard service luminances as taken from the CIBSE code

Task group	Typical	Standard service illuminance (lux)
Car parks		30–50
Circulation spaces	Corridors and passage ways	100–200
Vertical circulation spaces	Stairs, lifts escalators	150
No continuous work	Storage areas, plant rooms	150
Casual work		200
Rough work	Rough machining or assembly	300
Routine work	Offices, control rooms, medium machines and assembly	500
Demanding work	Deep plan, drawing or business machine offices. Inspection of medium machining	750
Fine work	Colour discrimination, textile processing, fine machining and assembly	1000
Very fine work	Hand engraving, inspection of fine machining or assembly	1500
Minute work	Inspection of very fine assembly	3000

BS 8300 shows the effectiveness of information on the use of a building is determined by:

- The location, accessibility, layout and height of signs
- The size of lettering, symbols and their reading distances
- The use of tactile letters and symbols
- Colour/luminance contrast and lighting
- The finished surfaces of materials used for signs and symbols
- The simultaneous use of audible cues
- Integration with any other communication systems.

Signage, in addition to being well designed, must be well positioned and illuminated. Poor surface finish, for instance a reflective surface, can completely negate the effectiveness of a sign, if people cannot read it because of the reflections.

The proliferation of signs can be as bad as the omission of information. Effective building design can minimise the need for excessive signage. Where they are needed they should be clear, well positioned, well illuminated with easily discernable text and/or figures. Good signage is vital for many people. This includes people with hearing impairments who will often not ask for directions as they expect not to be able to hear the answer. They are important to people with learning difficulties and short-term memory loss who may not be able to retain information given. Contrasting, obsolete, or conflicting information is not helpful.

The extracts in Figure 4.11 taken from 'The curious incident of the dog in the night-time' by Mark Haddon illustrates this well.

Further information and illustrations are given in the Appendix of this book, Workshops: some commonly queried scenarios, Scenario B. Query regarding Colour Contrast and Fonts.

Tactile signs and symbols for reading by people with visual impairments should be used for directional signs and signs identifying functions or activities in the building. The recommendations in BS 8300 are for:

- Embossed letters (rather than indented or engraved)
- Sans serif type face
- Depth of 1.25 ± 0.25 mm
- A stroke of 1.75 ± 0.25 mm
- Edges slightly rounded but not half-round in section
- Letter height between 15 and 50 mm.

Embossed letters are easier to read than indented or engraved letters, especially if their leading edges (left and upper) are as sharp and as well defined as possible. Also, over time indented letters tend to fill with dirt, polish etc., which renders them difficult to read.

Where Braille is used the recommendations are for the following:

- Grade 1 Braille should be used for single word signs

- Grade 2 contracted Braille should be used to reduce the length of multi-word signs
- Where Braille forms part of a sign, a marker (notch) should be located at the left-hand edge of the sign to help locate the Braille message.

The signs said

Sweet Pastries **Heathrow Airport Check-In Here** *Bagel Factory* **EAT** ***excellence and taste*** **YO!** sushi **Stationlink** Buses **W H Smith** MEZZANINE **Heathrow Express** Clinique First Class Lounge FULLERS **easyCar.com** ***The Mad Bishop*** **and Bear Public House** Fuller's London Pride Dixons **Our Price** Paddington Bear at Paddington Station **Tickets** Taxis **Toilets** First Aid **Eastbourne Terrace** ▇ing-ton Way Out **Praed Street The Lawn** Q Here Please Upper Crust Sainsbury's **Local** **Information** GREAT WESTERN FIRST Ⓟ Position Closed **Closed** Position Closed Sock Shop Fast Ticket Point **Millie's Cookies**

But after a few seconds they looked like this

Sweathr▇owOAirpheck-lagtoryEAenceandtaste YO! suuSetHeesortCWHSmithEANEINStatnH✳ioeadBho athrnieFirlassLoULLERnreHeBSeasyCar.comTheMpanard BebleFuler'sLonPrndoidePaiesstrDzzixonsOurisPPurdEboi ceicHousPatCngtoneaswatPoagtonTetsTaelFac Toil eddistsFirs—ta BungfeFi5us✱✖HPDNLeTerrace▇ ▇ingtonWastaySt atio nllnkOutClosed& qed3iniBr1uowo[CliPraicxiskedPointDrStreetTheLy uawHea rCrustMuflyBakl6dETonCloseexcel letoxpressnQinrePlek4shSaisesUp ← pensburiy'sLcidSo ktickmationREATM+ASTERCookiesWESTEfinsCojRN 2FningSTanl⑥RSTⓅP0allnforositioNCH✳EnSTAYATS 3hopFastPositdPeniesPloNla8⑨④tfoe9s

Figure 4.11 Illustration of confusing effects of excessive signage (Haddon, 2003)

4.6 Facilities in buildings other than dwellings

4.6.1 Objectives

This section includes the functions and facilities inside a building that people need access to and to use. One of the thrusts of the recommendations in the Approved Document is for all people to be able to participate in the proceedings, for example at lectures, conferences, theatre production or film showing, sports facilities and stadia, exhibitions and displays, refreshment facilities, sleeping facilities etc. Participation would include people as spectators, deliverers (e.g. actors, lecturers), staff and audience participants.

Facilities such as switches, outlets and controls are common to all buildings and need to be used by all people; guidance on these is therefore included at the start of this section.

Sanitary facilities are covered in a separate section.

The design considerations should include for people with visual, mental and hearing impairments, as well as those with mobility or dexterity problems.

4.6.2 Switches, outlets and controls

The principles of design given are ease of operation, visibility, height and freedom of obstruction. All users should be able to locate a control, know what setting it is on, and use it without mistakenly changing its setting to an undesired one.

Plates should visually contrast from their surroundings and be easy to use with a minimum of manual dexterity.

The colours 'red', 'green', 'brown' and 'black' should not be used in combination to distinguish between 'off' and 'on'. Many people have difficulties distinguishing between reds and greens, some see them as shades of brown, and others see both of them as black.

Pictograms and embossed text are useful to indicate a switch or control's purpose.

Where there are multiple switches or controls, these should be well separated to avoid a person operating one in mistake for another.

Requirements for ADM with further comment are shown in Table 4.9.

4.6.3 Audience accommodation and general considerations

Facilities for audience accommodation can be divided into four categories:

(a) Lecture, conference, meetings, etc. where there can be expected some participation between, or no distinction between, deliverers and audience. People should be able to be seated or positioned appropriately, use the presentation facilities, hear, see and respond to the proceedings without obstruction or impediment. As well as acceptable access and accommodation for wheelchairs, hearing enhancement systems, good sight-lines, design and location of lecture equipment, and good lighting are all important.

(b) Entertainment facilities, for example theatres, concert halls, cinemas where seating tends to be more closely packed and fixed. As above, people should be able to be seated or positioned appropriately, hear, and see the proceedings without obstruction or impediment. Acceptable access and accommodation for wheelchairs is required with a choice of position; also hearing enhancement systems and good sight-lines are important.

(c) Sports facilities. There are two considerations here; people as spectators and people as participants. In this section of audience accommodation, only spectators are considered.

(d) Leisure and social is grouped with entertainment in the Approved Document however, leisure and social would include refreshment facilities, bars, clubs, etc. where seating tends not to be closely packed or regimented, nor fixed. Many considerations would be for the management but any fixed seating design, and counters may be designed at Building Regulation stage. These facilities can be found in any of the other types of buildings.

There are general considerations for audience seating which should be considered:

- People with mobility or sensory impairments may need to view or listen from a particular side, or sit at the front for lip-reading or sign interpretation. Wheelchair users may also have these requirements, but not necessarily. Therefore a choice of seating should be available to all.
- Wheelchair users will need spaces into which they can easily manoeuvre. The positions should not be segregated from other areas, should have as good a view as other positions, and as good audibility.
- People with an assistance dog will require a space for the dog to sit and lie down.
- Some people will have difficulty using seats with fixed arms.
- Some people of large stature may need extra leg-room.
- Seats should contrast visually with the surroundings.
- People with any other requirements may also have a requirement to seat towards the front or towards the rear because of particular sensory conditions.
- Wheelchair users may wish to seat with other wheelchair users.
- People with assistance dogs may wish to sit with other people with assistance dogs.

Table 4.9 Requirements for switches, controls and outlets in non-residential buildings

		Requirement	Comment
1	Floor set outlets		These will be difficult to use by some people
2	Front plates	Contrast visually with the background	
3	All switches and controls requiring precise hand movements	Set between 750 and 1200 mm above floor level	
4	All controls requiring close vision	Readable parts to be set between 1200 and 1400 mm above floor level	So that readings can be taken by a person sitting or standing
5	Wall mounted outlets other than those in (3) or (4) e.g.: Sockets TV points IT points Telephone points	Set between 400 and 1000 mm above floor level Located consistently in relation to doorways and room corners No closer than 350 mm to room corners	To be reachable and usable from a wheelchair or by a person with limited vertical movement or by a person with limited dexterity
6	Switched socket outlets	Indicate whether they are 'on'	Note the use of red, green, brown or black must be used with care and not used to contrast with each other
7	Simple push buttons controls requiring limited dexterity	Not more than 1200 mm above floor level	To be reachable and usable from a wheelchair

8	Light switches used by general public	Large push pads Align horizontally with door handles Set between 900 and 1100 mm above floor level	For ease of use when entering room
9		Where (8) cannot be provided, a lighting pullcord, with a bangle of 50 mm diameter, which contrast visually from background, set at a height 900 to 1100 mm above the floor To be visually distinguishable from emergency pullcord	
10	Switches for permanently wired appliances	Set between 400 and 1200 mm above floor level	Unless needed at a higher level
11	Emergency pullcords	Coloured red, with two red bangles of 50 mm diameter, one set at a height 800–1000 mm above the floor, the other 100 mm above the floor	Visually obvious. Able to be used with minimal manual dexterity Usable from wheelchair, a seat, or when lying on the floor
12	Operation of switches, outlets and controls	Does not need the use of two hands simultaneously	For people who have the use of only one hand Unless necessary for safety reasons
13	Mains and circuit isolator switches	Clearly indicate that they are ‘on’ or ‘off’	Health and safety Note the use of red, green, brown or black must be used with care and not used to contrast with each other

- People who either use a wheelchair or an assistance dog may:
  - be accompanying their children and want to sit with them; or
  - be a child, wanting to sit with their parents, friends and siblings; or
  - as an adult want to sit with other adults.

Arrangements should be flexible to allow these common situations. Looking at these example issues, some solutions could cater for others. For example, greater spacing between seats for a person of larger stature could provide sufficient space for an assistance dog. Individual removable seats could permit wheelchair users to sit a suitable distance from the front, with other wheelchair users, and/or next to a person using a conventional seat. Flexibility in location is important.

4.6.4 Recommendations for all audience seating

(1) The route to wheelchair spaces should be accessible by wheelchair users, i.e. level or ramped access, with a clear width of at least 900 mm.

(2) Stepped access routes should be provided with suitable fixed handrail and acceptable step dimensions.

(3) An adequate number of wheelchair spaces should be provided in relation to the seating capacity. These should be mixture of permanent spaces and removable seats as shown in Table 4.10.

(4) Some spaces provided in pairs. This could be two permanent spaces, or one permanent space and one removable seat. Standard seating should be provided on at least one side of these pairs.

(5) Where at least 3 permanent wheelchair spaces are provided they are positioned to give a range of views on both sides of, at the front of and at the back of the seating area.

(6) The clear space allowance required for a parked occupied wheelchair is 900 mm wide by 1,400 mm deep.

(7) The floor of all wheelchair spaces (permanent or created by removable space) must be horizontal.

(8) Seats with space for assistance dogs to rest in front of, or under, the owner's seat. There is no recommendation for the number of such seats.

(9) Seats with removable or lift-up arms at the ends of rows and next to wheelchair spaces.

(10) A hearing enhancement system for lecture/conference type facilities.

(11) Wheelchair access to a podium or stage provided for lecture/conference type facilities. This could be by means of a ramp or lifting platform.

Table 4.10 Provision of wheelchair spaces in audience seating

Seating capacity	Permanent wheelchair spaces (min number, rounded up)	Removable seating for wheelchair spaces (min number)	Total minimum number of wheelchair spaces for ADM
Up to 100	1 (1% of total seating capacity)	5 (Remainder to make a total of 6 wheelchair spaces)	6
101–200	2 (1% of total seating capacity)	4 (Remainder to make a total of 6 wheelchair spaces)	6
Up to 600	1% of total seating capacity	Remainder to make a total of 6 wheelchair spaces	6
Over 600 but less than 10 000	1% of total seating capacity	Additional provision as required	1%
The following are taken from "Accessible Stadia", The Football Stadia Improvement Fund and The Football Licensing Authority (2003)			
10 000 to 20 000	100 plus 5 per 1000 above 10 000		n/a
20 000 to 40 000	150 plus 3 per 1000 above 20 000		n/a
40 000 or more	210 plus 2 per 1000 above 40 000		n/a

4.6.5 Sports facilities

The Approved Document does not give additional specific advice to that above for sports grounds but directs the designer to other publications such as the 'Guide to Safety at Sports Grounds', also known as The Green Guide, by the Department of National Heritage and the Scottish Office (1997), and "Accessible Stadia: A Good practice guide to the design of facilities to meet the needs of disabled spectators and other users" published by The Football Stadia Improvement Fund and The Football Licensing Authority (2003). The latter document is quite comprehensive discussing all aspects of accessibility including ticket purchase. It would be a useful tool to avoid discrimination under the DDA. The Green Guide, an extremely important document when published, does have some limitations. For example there are no recommendations for provisions for ambulant disabled people. "Accessible Stadia" uses The Green Guide recommendations extensively, having been updated where required and filled in some omissions. When designing sports facilities the designer should consult these documents as well as considering the recommendations in ADM.

4.6.6 Refreshment facilities

Cafes, restaurants, canteens, bars, public houses, coffee shops etc. can all form parts of larger buildings as well as being properties in their own right. People need to access these to purchase the refreshments as well as to work there. The areas should be designed so that they can be reached and used by all people independently or with companions.

Staff areas should be accessible to employees. This would include rest rooms, tea-making facilities, etc.

Facilities such as sanitary accommodation, public telephones, external terraces, beer gardens, should all be accessible.

Where self-service and waiter service is provided, all users should have access to both types of service. For example it would not be acceptable for premises to be designed so that a wheelchair user had to be seated at a table and use waiter service rather than seeing the food on display at the self-service counter.

Where changes in level are used to differentiate between different parts, for example in atmosphere, acceptability of children, or to allow smoking areas, all the different floor levels must be accessible.

- All users should have access to all parts of the facility.

- Counters etc should be usable by wheelchair users, so a part of the bar or serving counter should be permanently accessible at a level of not more than 850 mm above the floor (see also Figure 4.12).

- Parts of the counter should also be at a higher level for use by people standing. It is not comfortable for all people to have to bend down to collect food, write cheques etc.

- Where there is a transition from inside to outside, for example at external seating areas, the threshold should be level, or, if a raised threshold is unavoidable,

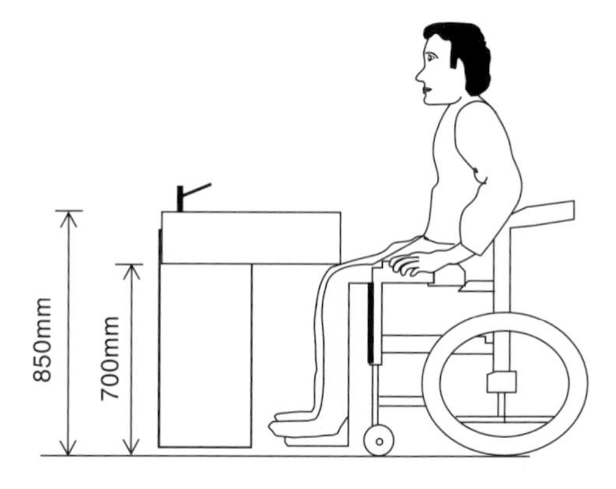

Figure 4.12 Counter heights for refreshment facilities

it should have a total height not more than 15 mm. Any upstands higher than 5 mm should be chamfered or rounded.

- In shared refreshment areas for example in rest rooms, or tea-making areas, the worktop should be at 850 mm above floor level with a clear space underneath minimum 700 mm above the floor. Taps should be of a lever action type or similar, and terminal fittings comply with the guidance relating to Schedule 2: Requirements for water fittings of the Water Supply (Water Fittings) Regulations 1999.

4.6.7 Sleeping accommodation

When sleeping accommodation is provided for a significant number of people, this should aim to be convenient for all people. The types of facilities discussed here includes in hotels, motels, and student accommodation.

Wheelchair users generally require more space than usually provided and are likely to need en-suite sanitary accommodation. This type of accommodation therefore has the most advice in ADM. Wheelchair accessible rooms are usable by the most numbers of people. In semi-permanently occupied property, where the accommodation resembles more a home, for example in student accommodation, it is useful to have a wheelchair accessible WC available for use by guests. Wheelchair users should be able to access all the facilities available in the building and wheelchair accessible bedrooms should be as advantageously placed as other bedrooms.

Because wheelchair users should be able to visit other people, all rooms including all bedrooms, should have doors wide enough for a wheelchair to get through, whether they are accessible bedrooms or not. The 300 mm access space is not necessary to the side of these doors as it is presumed that the occupier of the room being visited will open the door.

Entrance doors to wheelchair accessible bedrooms can have powered opening. This can avoid the need for the 300 mm access space needed to the leading edge side of the door for manual opening doors. For unlocking the door, devices such as electronic key cards and level action taps can be essential for people with limited manual dexterity.

The size of bedrooms, and space for and placing of furniture and fittings, can make a big difference in the convenience of sleeping accommodation. In particular people with mobility problems, including wheelchair users need to be able to access and use fitted wardrobes and shelving. Automatic curtain closing devices can also be helpful. Wheelchair accessible bedrooms should be sufficiently spacious for a user to be able to reach both sides of the bed, use all the facilities in the accommodation, including any sanitary accommodation and balconies, and operate switches and controls.

It is preferable to provide en-suite facilities for a wheelchair user. There should be equal numbers of en-suite shower rooms as en-suite bathrooms because showers are easier for some wheelchair users.

When providing wheelchair accessible bedrooms some interconnecting rooms are helpful; these can accommodate companions, or accompanying children, parents, etc.

4.6.8 Recommendations for all bedrooms

(1) The doors to all bedrooms to have a minimum clear width of 800 mm or wider to comply with Table 4.3, dependent on the corridor width.

(2) Any room numbers are shown with embossed characters.

(3) All bedrooms are to have a visual fire alarm warning as well as an audible one, in addition to other requirements of Part B of the Building Regulations.

(4) For wardrobes and other fittings that have swing doors, the doors should swing through 180°. This allows the greatest access to the interior.

(5) Handles on all hinged and sliding doors (to the structure or to fittings) should be easy to grip and operate and be easily distinguishable from their background.

(6) Openable windows and the window controls are to be positioned between 800 and 1000 mm above floor level. Windows should be able to be opened or closed using only one hand, without needing excessive strength or to use excessive force.

4.6.9 Additional recommendations for wheelchair accessible bedrooms

(7) One in every 20 bedrooms should be wheelchair accessible, as a minimum.

(8) Accessible bedrooms should be located on accessible corridors giving access to all the building's other facilities.

(9) There should be the same choice of location and standard of amenity as in other bedrooms.

(10) Bedroom entrance doors are to have:
- at least the minimum accessible width in relation to the corridor width,
- the door should not need excessive force to open it, i.e. a maximum of 20 N,
- there should be a 300 mm unobstructed space of 300 mm to the leading edge side of the door.
- all other relevant requirements of the door as an 'internal door'.

(11) The effective clear width of the door into the en-suite should be as Table 4.3, i.e. a minimum of 800 mm dependent on the space available to manoeuvre into the en-suite room.

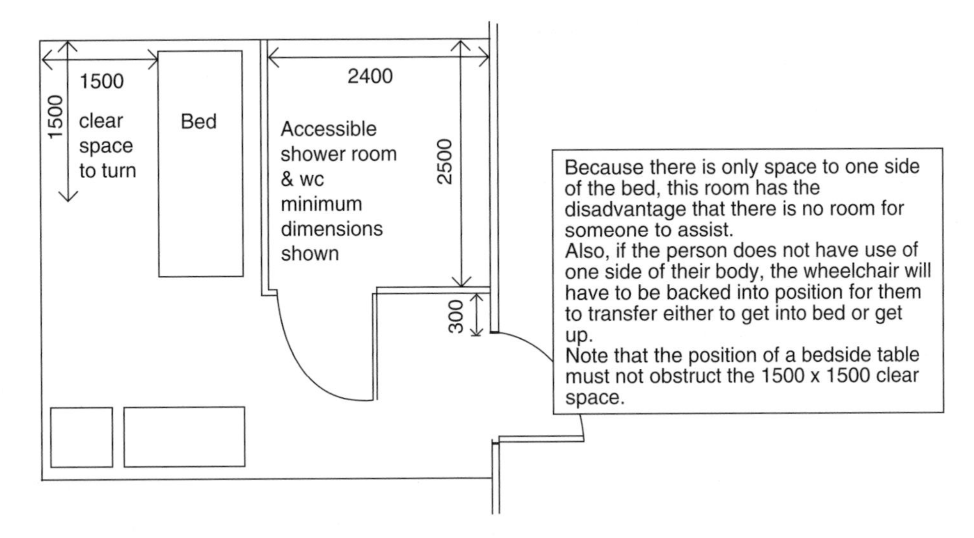

Figure 4.13 Minimum sized accessible bedroom with en-suite

(12) A clear width of 1500 mm is needed to one side of at least one bed in the room to allow for a wheelchair user to manoeuvre at the side of the bed and then to transfer into it (see Figure 4.13). A clear width of 700 mm to the other side allows for a carer to help the person into bed.

4.7 Sanitary accommodation in buildings other than dwellings

4.7.1 Principles and general design considerations for sanitary accommodation

These provisions relate to WCs, shower rooms, and changing facilities. The main principle and objective is that sanitary facilities are available for use and accessible by all people. This means that as well as people who are adult, able-bodied, need a minimum of time and space to use the facilities (i.e. people who do not have additional problems), other people who cannot walk so far or so quickly or who are necessarily encumbered by sticks, pushchairs, small children, wheelchairs etc. are able to reach the facilities in time. Other people have different issues to the number, positioning, and size of accommodation.

- People with visual disorders need to be able to locate the facilities without unusual difficulty. In multi-storey buildings, the consistent location of accommodation is helpful here, as well as tactile messaging.
- Similarly people with mental or memory impairments should not find the locations confusing.

- Small children need to be able to reach the WCs, urinals and wash basins.
- Once into the accommodation, people need to be able to switch on the lights if they are not permanently on. Pad light switches are easier than pullcords.
- They need to be able to get into the WC compartments and close the door without being obstructed by the WC and other items in there.
- Door furniture needs to be usable with minimal strength and dexterity.
- Taps should be similarly usable by everyone and not scald them by their hot surface or the temperature of the water.
- Storage cupboards should be provided for cleaners' equipment, etc so that the accessible compartments are not used as depository areas for mops, buckets and vacuums. The clear space for manoeuvrability should be permitted to remain clear.
- Doors should open if a person in the cubicle fall against them, and open outwards where possible or not encroach on the manoeuvrability space if inward opening.

The design considerations offered by the Approved Document generally for sanitary accommodation include:

(a) Taps that can be used with a closed fist – a lever design being the most common.
(b) Fittings that do not permit burning or scalding (i.e. guidance note G18.5 relating to Schedule 2: Requirements for water fittings of the Water Supply [Water Fittings] Regulations 1999).
(c) Door furniture (handles, latches, locks, bolts etc.) that contrasts visually with the door surface, and is usable with one hand and a closed fist. BS 8300 also offers that door furniture is positioned in the zone 900–1100 mm above the floor and preferably at 1000 mm above floor level. BS 8300 also recommends that handles are set minimum 54 mm from the edge of the door, to avoid scraping the knuckles.
(d) Privacy bolts should be light-actioned so that they are able to be used with a minimum of strength and manual dexterity.
(e) Any doors with self-closers should not need a force greater than 20 N to open them.
(f) Emergency release mechanisms should be provided to doors, so that they can be opened outwards from outside in an emergency.
(g) Doors should not obstruct emergency escape routes.
(h) Fire alarms provided should be visual as well as audible
(i) Any emergency assistance alarm system complies with the requirements in Table 4.11, which gives recommendations from ADM and further clarification and advice from BS 8300.
(j) Lighting controls to be as in Table 4.9.
(k) Hot surfaces or heat emitters to be provided with guards or have their surface temperature kept below 43°C to avoid a person burning themselves. Some

Table 4.11 Emergency assistance alarm system requirements

	Requirement	Comment
Emergency assistance pullcord		Reachable from a wheelchair Reachable from the WC Reachable from the shower/changing seat
	Coloured red, with two red bangles of 50 mm diameter, one set at a height 800–1000 mm above the floor, the other 100 mm above the floor	Visually obvious Able to be used with minimal manual dexterity Usable from wheelchair, seat, or when lying on the floor
Indicators	Inside compartment Visual and audible	To confirm to the person that the emergency call has been received
	External to compartment Located so that it is seen and heard by people able to give assistance and indicates where help is required	So that help is achieved
	Not able to be confused visually or audibly with the fire alarm	
Reset control button	To the side of the WC, shower seat or changing seat Between 800 and 1000 mm above floor level	Reachable from a wheelchair Reachable from the WC Reachable from the shower/changing seat
	Clearly marked as such	

people have conditions where they are not alerted to heat signals through their skin and are therefore prone to burns, for example, multiple sclerosis (MS) sufferers can sometimes have this symptom, and they should be therefore permitted protection from hot surfaces.

(l) Appliances, grab handles, controls etc. should all contrast visually with their backgrounds. There should also be visual contrast between the walls and floor. White appliances against white walls and floors would obviously be difficult to locate for some visually impaired people. Contrast can be provided in different ways, for example if white tiled walls and floors are required, then a dark blue skirting could be used to delineate the extent of the floor surface.

4.7.2 Provision of general toilet facilities and design considerations

Toilet facilities need to be suitable for all people who use the toilets. This includes parents with children, people with luggage, ambulant disabled people, wheelchair users, parents needing baby-changing facilities, carers needing adult-changing facilities.

Table 4.12 Provision of toilet accommodation

	Provision to be	Additional comment
Space for only 1 toilet in building	Wheelchair accessible unisex type 2000 mm wide minimum With a standing height wash basin (i.e. height to rim from floor is between 780 and 800 mm) in addition to the finger rinse basin associated with the WC	A greater width than the minimum is required to accommodate the additional wash basin This standing height basin can be used by people with chronic back problems for whom bending is painfully difficult
At each location of toilet provision for public or staff	At least one separate wheelchair accessible unisex toilet	
In separate-sex toilet facilities	At least one accessible toilet for ambulant disabled people	This is a requirement in all single sex toilet facilities, so if there is room for just one male and one female, both these should be of ambulant disabled design (and have a door capable of being opened outwards)
Where there are four or more cubicles in separate-sex facilities	At least one accessible toilet for ambulant disabled people (as above), and at least one other cubicle to be extra large (see 'enlarged cubicle' below). The remaining two or more cubicles can be 'standard'. These will be complemented by a fully accessible cubicle nearby	For people who need extra space (e.g. people with small children, people with luggage etc.)
WC cubicles within single-sex toilet washrooms	450 mm diameter manoeuvring space is maintained between the swing of the door, the WC pan and the side wall inside the cubicle.	This allows people to close the door without standing on the WC. It also permits baggage to be taken into the cubicle which is better for security

WC cubicles for ambulant disabled people	800 mm wide minimum 750 mm length of activity space in front of WC pan clear of door swings Door to preferably open out with horizontal closing bar fitted to inside face 500 mm long horizontal grab rails at 680 mm above floor, set to extend minimum 200 mm beyond toilet pan WC seat at 480 mm above floor WC to comply with the key dimensions given in BS 5503-3 and 5504-4 Floors to be slip resistant See also Figure 4.14	Ambulant disabled people may need more space to manoeuvre due to impaired leg movement, walking aids, crutches etc. Some people have difficulty using standard height WCs and carry a toilet seat riser. The key dimensions of the WC should be such that these will fit securely.
An enlarged cubicle	1200 mm wide minimum A horizontal grab bar adjacent to the WC A vertical grab bar on the rear wall A shelf A fold-down changing table	For people who need extra space (e.g. people with small children or babies, people with luggage, etc.)
Any wheelchair accessible washroom	A wheelchair accessible compartment (where provided) has the same layout as in Figure 4.15 At least one wash basin with the rim set between 720 and 740 mm above the floor For men, at least one urinal with the rim set at 380 mm above the floor Either side of this urinal should be two vertical 600 mm long grab handles centred 1100 mm above the floor See Figure 4.14(c)	The low basin and urinal provides for children and wheelchair users

Parents could be male or female. People accompanying wheelchair users could be male or female.

A solution for one set of people may not be acceptable for another. For example, wheelchair accessible cubicles provided in single sex facilities can be awkward for a wife who needs to help her husband to use the lavatory. Do they both go onto the male facilities or the female? For this reason wheelchair accessible unisex toilets should always be provided in addition to any accessible single sex facilities.

An accessible cubicle in single sex facilities can be helpful to people with small children or luggage, and ambulant disabled people as well as wheelchair users.

The ADM states that the requirements of Regulations M1 and M3 will be satisfied by the provisions in Table 4.12.

4.7.3 Wheelchair accessible unisex toilets

These are required wherever there is provision of toilet facilities in the building. Where there is room for only one WC, it must be a wheelchair accessible unisex toilet.

The relationships between the different provisions required i.e. the door, manoeuvring space, finger rinse basin, WC, handrails, cords, mirror etc. are all vital to ensure that a person who uses a wheelchair is able to use the facilities. Space must not be obstructed with other things (e.g. stools, bins, cleaners' equipment, chemical sanitary disposal unit etc.) therefore space should be provided for these elsewhere within the cubicle or without.

A person should be able to:

- enter the accommodation
- close and lock the door
- position the wheelchair close to the WC pan
- transfer across to sit on the toilet by the means suitable for them (from the left, the right, or the front with or without assistance)
- use the support and grab rails as they need to
- wash and dry their hands while seated on the WC
- transfer back again, using the support as needed
- tidy themselves, using the mirror set at an appropriate height
- manoeuvre the wheelchair so as to face the door
- and leave with ease.

One unisex wheelchair accessible toilet should be located as close to the entrance and waiting areas as reasonable but not in any location where privacy or dignity could be compromised.

Other unisex wheelchair accessible toilet should be located via accessible direct and unobstructed routes in a consistent manner throughout the building so that they are easily found. There should be equal numbers of left-and right-handed transfer WCs on alternate floors.

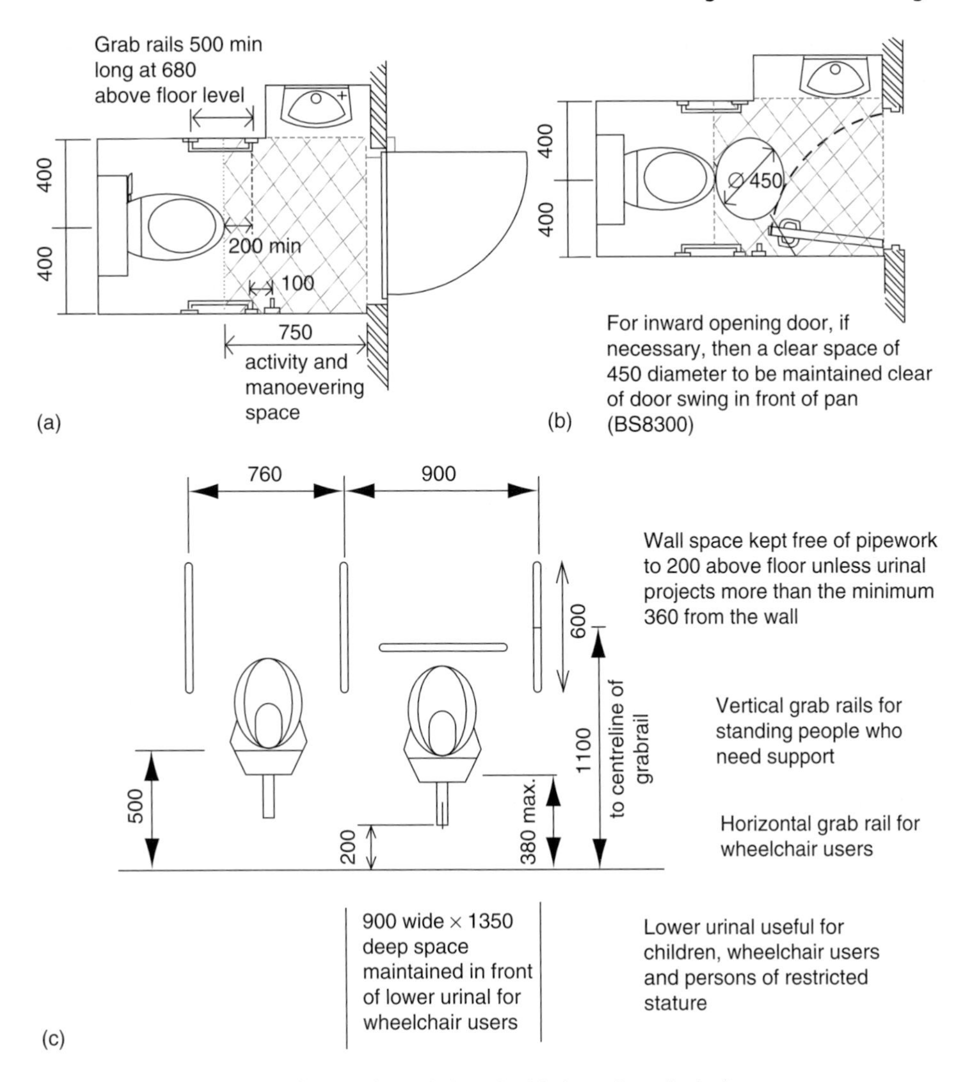

Figure 4.14 Key dimensions for WCs for ambulant disabled people and urinals

Users should not have to travel more than 40 m on the same floor. An exception is given for this if it can be shown via the Access statement submitted with the application for Building Regulation approval that although the distance is greater than 40 m, it is on an obstructed circulation route, for example, with doors held open on automatic door release mechanisms.

Where the WC is on a different floor and there is an accessible lift giving access between floors, users should not have to travel more than 40 m combined horizontal distance. Where there is not a passenger lift but a platform lift, users should not need to travel more than one storey. This is because of the time factor involved with travelling more than one floor by platform lift.

See Figure 4.15 for the minimum dimensions involved with a standard unisex wheelchair accessible toilet. General provisions, such as emergency assistance alarms, as detailed in Table 4.11, also should be observed.

Any heat emitters (radiators etc.) in the accommodation must not obstruct the clear manoeuvring space of at least 1500 × 1500 mm, nor the space beside the WC used for transferring onto the WC.

The flushing mechanism for the WC should be located on the open side of the WC however they operate, so that the toilet can be flushed when the person is back in their wheelchair.

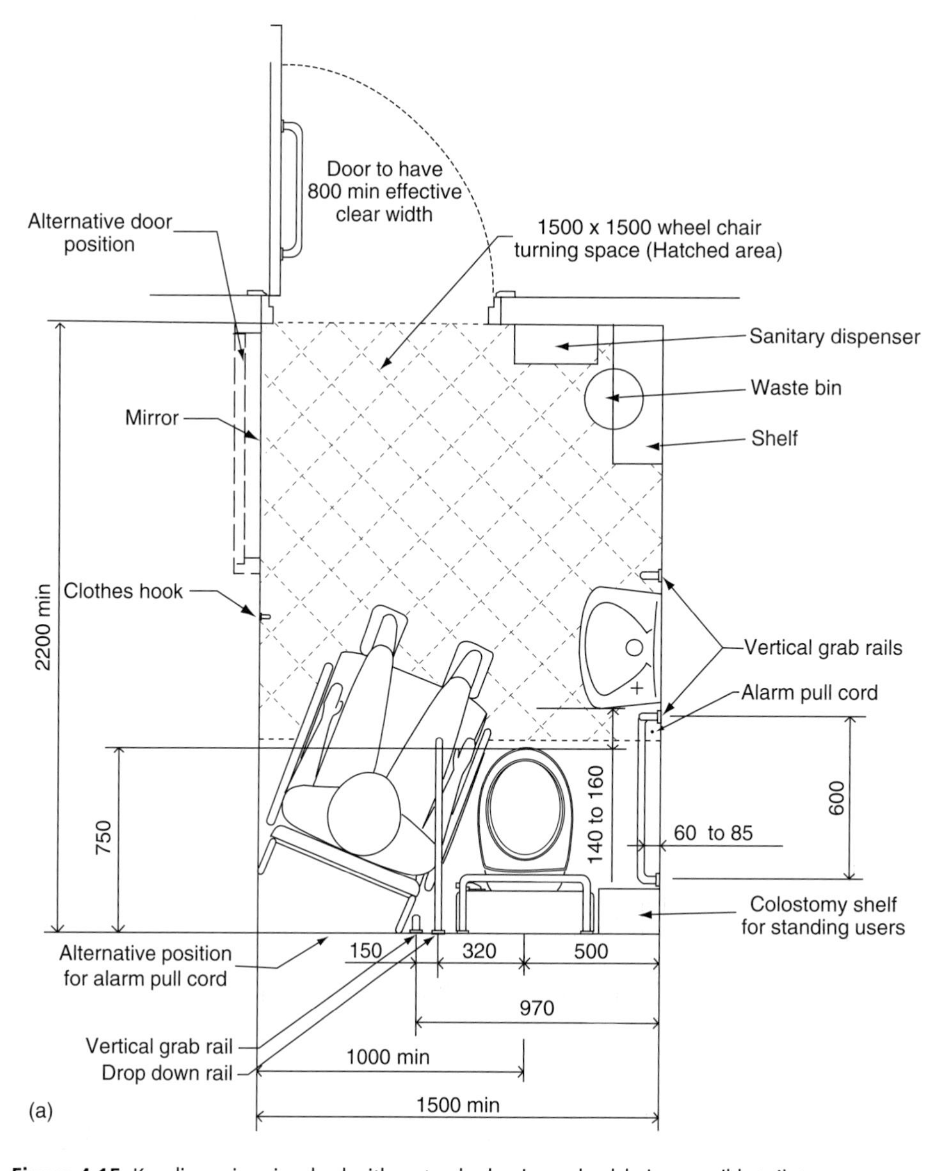

Figure 4.15 Key dimensions involved with a standard unisex wheelchair accessible toilet

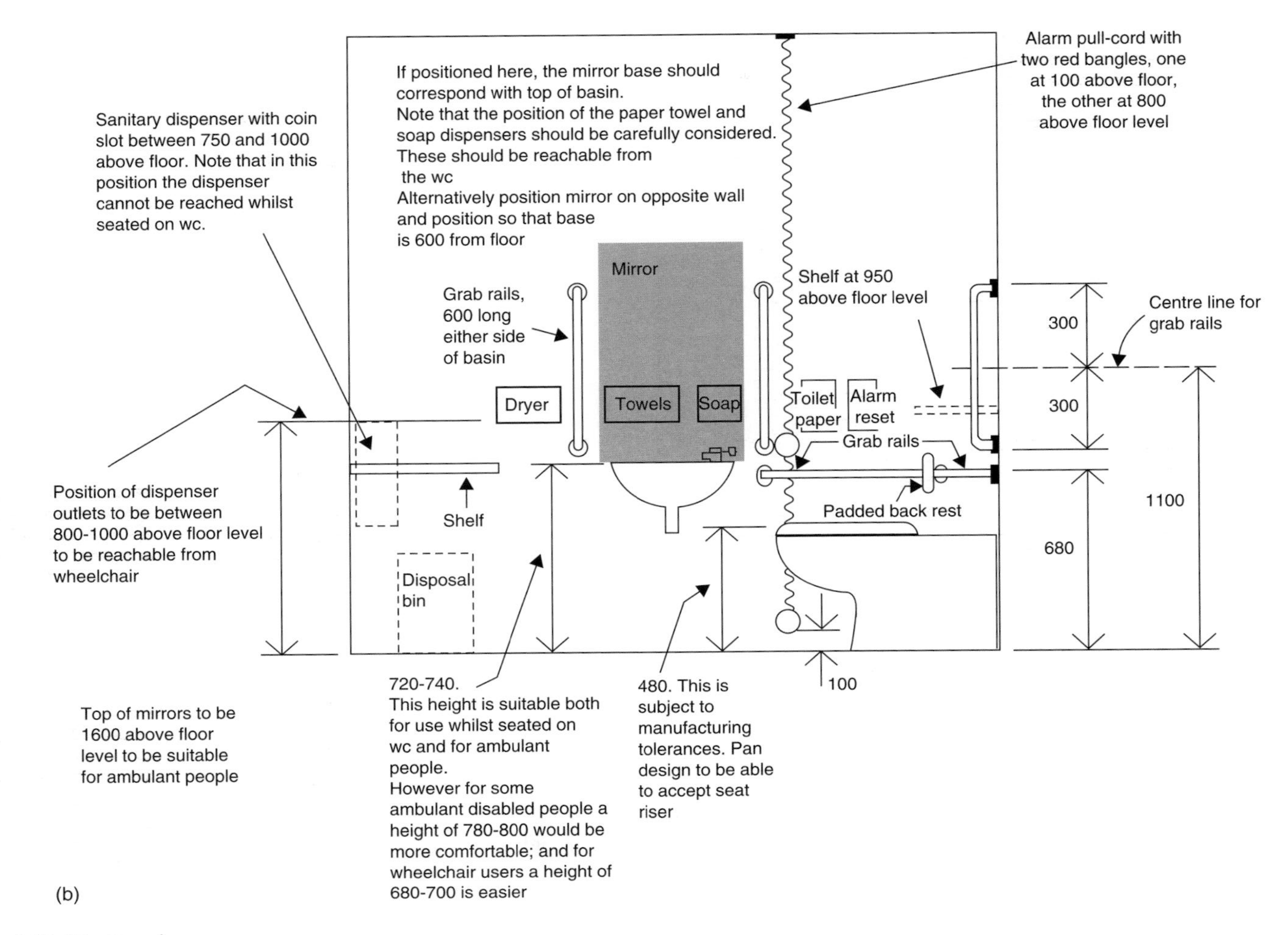

Figure 4.15 *(Continued)*

The WC pan must conform to the key dimensions given in BS 5503-3 and 5504-4, so that it can accommodate a toilet seat riser. Some people have difficulty using standard height WCs and carry a toilet seat riser. These are like a thick toilet seat which can be easily fitted onto and secured to a WC pan raising the seating level perhaps 75 mm or 100 mm, making it easier for a person to lower and raise themselves from the toilet. The key dimensions of the WC should be such that these will fit securely.

4.7.4 Changing facilities and showers accessible with wheelchair

Types of buildings associated with changing room and/ or shower facilities would be clothes shops, leisure complexes, sports facilities, and swimming pools. Some workplaces also provide shower and changing facilities for staff. These changing facilities and showers should be wheelchair accessible.

In larger buildings or complexes such as retail parks and large sports centres there should be at least unisex wheelchair accessible toilet with an adult changing table.

Different considerations need to be taken into account here, such as privacy where required, and space for a helper. 'Wet' and 'dry' areas need to be segregated so that people do not slip and wet surfaces to be slip resistant (see also Table 4.9) and self draining.

Manoeuvring space of 1500 mm deep in front of lockers should be provided in accessible self-contained and communal areas.

Aids such as wall-mounted drop down support rails and wall-mounted slip-resistant tip-up seats are required. The seats should not be spring loaded.

Emergency assistance alarm system should be provided with the pullcords reachable for a wheelchair space, tip-up seat or floor.

The only wheelchair accessible WC facilities should not be in 'wet' areas.

For wheelchair accessible spaces a choice of left or right transfer layouts is required where there is more than one individual compartment.

Shower and other terminal water fittings are to comply with guidance note G18.5 relating to Schedule 2: Requirements for water fittings of the Water Supply (Water Fittings) Regulations 1999. Controls should be easy to use with minimum manual dexterity and easy to understand, and fitted between 750 and 1000 mm in wheelchair accessible communal facilities.

Suitable limb storage is helpful for amputees.

Self-contained accessible units in addition to communal are required for sports facilities. These units require a minimum floor space of 2200 × 2000 mm.

Self-explanatory layouts are given in the Approved Document for self-contained changing rooms and shower rooms incorporating a WC. As in wheelchair accessible WC accommodation, space for manoeuvring of 1500 × 1500 mm clear of the facilities and obstructions is required together with transfer space, and controls reachable from the WC, wheelchair, seat provision or floor in an emergency. Grab rails need to be provided for assistance and clothes hooks reachable from a seated position are helpful.

4.7.5 Wheelchair accessible bathrooms

The guidance here includes accommodation for washing and bathing and would include bath or shower facilities. It differs from the previous section as it relates more to living accommodation. The type of building referred to would include hotels, motels, student accommodation and relative accommodation in hospitals. Other advice has been incorporated in Section 4.6 "Facilities In Buildings Other Than Dwellings", *Sleeping accommodation* onwards.

Minimum floor space requirements for:

- a shower room with corner WC would be 2400 × 2500 mm
- a bathroom with a corner WC would be 2500 × 2700 mm
- both with the clear space for wheelchair turning of 1500 × 1500 mm.

Where there is more than one provision there should be a choice of left or right transfer.

The floor of the bathroom should be slip resistant whether wet or dry (see Table 4.12).

A bath should be provided with a transfer seat, 400 mm deep and equal to the width of the bath.

Normal measures such as the door opening outwards, with a horizontal closing bar and emergency assistance pullcords and alarm system should also be provided. See Section 4.7.1 'Principles and general design considerations for sanitary accommodation'.

4.7.6 Fire alarm provision in sanitary accommodation

If the building has a fire alarm system then the sanitary accommodation must also be covered by that system. People can obviously take longer to be able to leave this area than other areas of the building.

People in any of the WC cubicles (standard, enlarged, designed for ambulant disabled people, or for wheelchair users) and any of the other provided sanitary facilities, therefore need to have some indication to be aware that the fire alarm is activated. This indication must be audible and visible. People with hearing impairments will not necessarily be using any particular type or size of cubicle, and therefore there needs to be visual indication in every cubicle. There does not need to be an audible device in every cubicle, as long as the sounders are acceptably audible throughout and in every part of the sanitary facility.

5

Dwellings

5.1 Introduction

5.1.1 The main provisions

Provision should be made for suitable access to, and into buildings including the use of their facilities. This provision does not apply to extensions or material alterations of dwellings. In order to achieve access to dwellings, consideration should be made for:

- A suitable approach from the boundary of the site to the entrance storey or the principal storey of the dwelling.
- Access into the building.
- Access within the building.
- The use of facilities.
- Sanitary conveniences in the entrance storey or, where the entrance storey does not contain habitable rooms, in the entrance storey or the principal storey.

It should be noted that the provision of access and facilities in dwellings is to enable people of all ages, and abilities to visit new dwellings and use the principal storey. It will also mean that as an occupant ages, or if they start to have a mobility impairment, they can cope better with impairments and remain in their home longer than would otherwise be the case. The provisions are not intended to facilitate fully independent living for all disabled people as each person's needs will differ from another's, but the requirements will reduce the number of common problems due to dwelling design encountered by people of all ages.

5.1.2 Interpretation

The following definitions of terms used in Approved Document M (ADM) apply only to the provisions concerning dwellings:

Clear opening width – For dwellings, the clear opening width of a door is taken from the face of the door stop on the latch side to the face of the door when opened at 90°.

Common – serving more than one dwelling.

Dwellings – This term means a house or a flat or maisonette. It does not include hotel accommodation or motels; this type of accommodation is covered under 'Buildings other than dwellings'. Purpose built flats used as student accommodation are regarded as a mixture of both, with general provisions as 'Dwellings', but in respect of space requirements and internal facilities they are to be treated as hotel/motel accommodation in 'Buildings other than dwellings'. One reason for this is that, similar to hotels, the provider of the accommodation may have little knowledge of the requirements of the occupier, and more particularly the occupier may have little choice in their accommodation. When renting or buying a house or flat, however, the occupier usually has the choice to accept the accommodation or reject it if it does not meet their needs. Therefore, refer to the clauses discussing 4.17–4.24 of the ADM for advice regarding space requirements and internal facilities for student accommodation.

Entrance storey – This is defined in the Regulations for Requirement M4, i.e. dwelling only as meaning the storey which contains the principal entrance.

Habitable room – This term is used for defining the principal storey of the dwelling. It means a room used, or intended to be used, for dwelling purposes and includes a kitchen, but not a bathroom or utility room.

Maisonette – A self contained dwelling, but not a dwelling house, which occupies more than one storey in a building.

Point of access – The point at which a person visiting a dwelling would normally alight from a vehicle which may be within or outside the plot, prior to approaching the dwelling.

Principal entrance – The entrance which a visitor not familiar with the dwelling would normally expect to approach, or the common entrance to a block of flats.

Principal storey – This is defined in the Regulations for Requirement M4, only, as meaning the storey nearest to the entrance storey which contains a habitable room, or if there are two such storeys equally near, either such storey. For example, one may enter the dwelling by the front door on a level "A" which contains only an entrance hall, and the stairs to an upper level "B" where, say, the kitchen and dining room are. Level A is the entrance storey and level B is the principal storey.

Plot gradient – The gradient measured between the finished floor level of the dwelling and the point of access.

Steeply sloping plot – A plot gradient of more than 1 in 15.

5.1.3 Objectives of the provisions

For dwellings, the requirements are for people to be able to access the dwelling, use the facilities, and have sanitary provision in the entrance storey if it contains a habitable room. This is to enable the occupant to have guests who may use a wheelchair.

If the entrance storey does not contain a habitable room and people have to move to a different storey to reach a habitable room anyway, then the sanitary conveniences may either be on the entrance storey, or on the different storey. The primary objective, as stated in the Approved Document is to provide a WC in the entrance storey of the dwelling and to locate it so that people do not need to negotiate a stair to reach it from the habitable rooms in that storey. The implication is that the WC will be accessible to wheelchair users.

This raises the question of steeply sloping sites and dwelling designs which have no habitable rooms on the ground level (entrance storey). The Approved Document discusses steeply sloping sites, suggesting that where the site slopes too greatly for reasonable provision of a ramp, suitable steps and stairways which are usable by ambulant disabled people and people with visual disorders may be sufficient. It should be borne in mind by the designer however that this will greatly inconvenience people with small children and pushchairs, and elderly people.

For flats, the Approved Document states that the objective should be to make reasonable provision for disabled people to visit occupants who live on any storey, and states that a lift would be the most suitable means of achieving this. However, it does go on to state that a lift may not always be provided. Again where a lift is not provided the stair should be suitable for ambulant disabled people and for people with visual disorders. Once again this would not usually be convenient for elderly people or for people with small children and pushchairs. However, this acknowledgement in the Approved Document for the non-provision of a lift might be used as support for a stairs only situation, where this is seen to be reasonable.

Inside dwellings (houses or flats), the requirement is for WC provision on the principal storey where there is no habitable room on the entrance storey. This would seem to suggest that the dwelling could be designed without habitable rooms, and therefore without WC provision on the ground floor. However, this would go against the requirements of M1 which requires access to use the building and its facilities. Therefore, in this situation, provision would have to be made for access to the principal storey. For flats, the provision can be in the entrance storey of the building, rather than the actual flat.

Because the definition of dwelling includes a flat as well as a house, there can be confusion while trying to interpret the recommendations for flats; many of the recommendations seem to be more relevant to a house than a flat. For example, the requirement for an accessible WC (as in Figure 5.7) in a flat on the second floor of a building which is not provided with a lift would seem unreasonable.

5.2 Getting to and into the building

5.2.1 Approach to the dwelling

In general terms, the Approved Document states that people should be able to approach and gain access to a dwelling from the point of alighting from a vehicle which may be outside the dwelling's plot. In most cases, it should be possible to provide a safe and convenient level or ramped approach, thereby enabling small children, people with prams and pushchairs, wheelchairs, and walking trolleys to easily gain access to the dwelling. A level approach is one where the route is flatter than 1 in 20. A ramped approach is one where the route is between 1 in 20 and 1 in 15.

The important considerations of a ramped access are discussed in BS8300 (BSI, 2001), as being the gradient of the ramp and the distances between the individual flights of the ramp. When the gradient is too steep or the flight is too long, people using the ramp, particularly those in a wheelchair or with reduced mobility or health, may not have sufficient strength to move themselves up the slope. A companion pushing a wheelchair will also face similar effects. If the gradient is too steep there are also the dangers when moving downhill of a person with mobility problems, such as an elderly person slipping, or of a wheelchair user falling out forwards. A wheelchair can tip over backwards when going uphill. Cross gradients add even more problems.

The Approved Document does not recommend the provision of handrails for a ramp to a dwelling, but for ambulantly disabled people, especially on long or steep ramps, and in slippery conditions, handrails such as provided for steps or for ramps to non-dwellings would be helpful.

There will be situations on steeply sloping sites where a ramped access would be greater than 1 in 15, when steps suitable for ambulant disabled people will have to be provided instead of a ramp. Steps without an associated ramp should only occur in unusual circumstances. When a ramp would be quite steep, but less than 1 in 15, the provision of suitable steps as well as a ramp would be a great advantage to many people.

The choice of a suitable approach to the dwelling from the point of access to the plot, will be influenced by the topography and available area of the plot. Location and arrangement of the dwelling on the site and the plot is a matter for Planning consideration and, therefore, the requirements of the planning authority should be taken into consideration. Account should also be taken of any listed building environs or conservation area requirements. Developers are advised to discuss the access requirements of Part M with their local planning authority, in conjunction with their building control supervisor (local authority or approved inspector) at an early stage in the design process, to avoid later conflicts. This may be particularly relevant if using a type approval system.

As the Approved Document suggests that reasonable access would not necessarily be from the boundary of the plot but from the point of alighting from a vehicle, it may be possible to reduce the effect of a steeply sloping plot by means of a suitable driveway. This could allow for the parking space within the plot boundary to be at a high level to permit a level or ramped approach from the parking space to the dwelling. This would of course not solve the problem for people for not travelling by vehicle, particularly those without the use of a vehicle and for short journeys, for example, to and from bus stops. However, the Approved Document, with its definition of "point of access" suggests that this may satisfy the requirements of M1.

The surface material of the approach to the dwelling should be firm enough to support a wheelchair and user, smooth enough to allow easy manoeuvre by all users, especially those of wheelchairs, pushchairs, trolleys, crutches and sticks, and slip-resistant. Loose material such as gravel or shingles is not suitable, nor are cobbles, bare earth, sand or similar materials.

The approach should be sufficiently wide, at least 900 mm, in addition to any parking space to allow safe and convenient passage. The approach should also not have any crossfalls greater than 1 in 40, as these can unbalance people.

In practical design terms the provisions can be summarised in the following paragraphs.

Provide a suitable approach:

- From a reasonably level point of access to the dwelling entrance (i.e. from the vehicle parking position to the principal entrance, or a suitable alternative if necessary).
- With crossfalls which do not exceed 1 in 40.
- Which may consist, in whole or in part, a vehicle driveway.

A level approach will have:

- A gradient not exceeding 1 in 20
- A firm and even surface
- A minimum width of 900 mm.

Where the overall plot gradient exceeds 1 in 20 but not 1 in 15, a ramped approach will be needed which should have:

- A firm and even surface.
- A minimum unobstructed width of 900 mm.
- Individual ramp flights no longer than 10 m in length with a maximum gradient of 1 in 15.
- Where an individual ramp flight does not exceed 5 m in length, a gradient not exceeding 1 in 12 is allowed.
- Top, bottom and, if necessary, intermediate landings at least 1200 mm long are required, clear of any door or gate swinging across it.

A stepped approach will be needed under the circumstances of an overall plot gradient exceeding 1 in 15. In this case the stepped approach should have:

- Minimum unobstructed flight widths of 900 mm.
- A maximum rise of 1800 mm between flights.

- Top, bottom and, if necessary, intermediate landings at least 900 mm long are required, clear of any door or gate swinging across it.
- A suitable tread profile as illustrated in Figure 5.4.
- Uniform risers between 75 and 150 mm high and goings of at least 280 mm.
- Tapered treads are permitted where the going, measured at a point 270 mm from the "inside" of the tread is not less than 280 mm.
- Where the flight consists of three or more risers, a suitable handrail should be provided on one side. The handrail should be continuous, with a profile which can be gripped, be positioned between 850 mm and 1000 mm above the pitch-line of the steps, and project at least 300 mm beyond the top and bottom nosings (see Figure 5.5). It should be noted that a handrail on only one side can make it awkward for people who have the full use of only one hand or arm, or when helping small children to use the steps.

Where a driveway provides the means of approach to the principal entrance, this should be such that the alighting point provides a route to the principal (or other suitable) entrance:

- Past any parked cars
- In accordance with the provisions described above as appropriate.

5.2.2 Access into the dwelling – level access

In general the access into a dwelling or block of flats from the outside should be provided with an accessible threshold, irrespective of whether the approach is level, ramped or stepped. Exceptionally, if the approach has to be stepped, and for practical reasons a step into the dwelling is unavoidable, it should not exceed 150 mm in height.

Level thresholds need to be designed carefully to avoid problems of water ingress, continuation of perimeter floor slab insulation if provided, and continuation of any gas barrier for instance, on contaminated sites if the area is not to admit water, moisture or contaminants, and not be at risk of condensation. Current observations of site work suggests that these issues are not given enough consideration by many designers and site staff. Some guidance on level thresholds is given in the publication 'Robust Standard Construction' (DTLR, 2001) and in the publication 'Accessible thresholds in new housing: guidance for house builders and designers' (TSO, 1999). Refer also to the requirements of Regulation C1, site preparation and resistance to contaminants, and to C2, resistance to moisture (see Figure 5.1).

Approved Document C states that an accessible threshold will meet the requirements of Regulation C1 and C2, if the external landing is laid to a fall of between 1 in 40 and 1 in 60 in a single direction away from a doorway and that the sill leading up to the door threshold has a maximum slope of 15°.

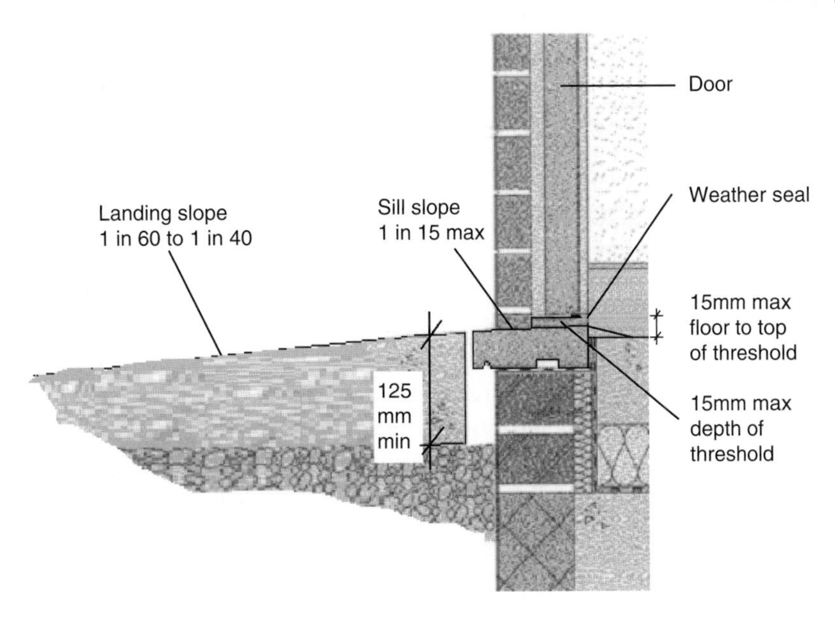

Figure 5.1 Robust construction details for level threshold

BS 8300 (BSI, 2001) gives additional information such that the entrance threshold should either be level, or, if a raised threshold is unavoidable, of a height no more than 15 mm and any threshold in excess of 5 mm should have a rounded or chamfered edge. Any upstands impede access as any small variation from otherwise level routes can make passage impossible or even dangerous. There can be trip hazards as well as obstructions for wheels. People pushing wheelchairs, pushchairs or trolleys also find sharp upstands difficult to manage.

5.2.3 Access into the dwelling – entrance doors

On reaching the building with an unobstructive threshold, the person should be able to easily negotiate the entrance door. The entrance door to any individual dwelling or into a block of flats from the outside should be wide enough to accommodate the person. Taking a wheelchair user as a basis for the widest expected need, entrance doors should have a minimum clear opening width (see section 5.1.2) of 775 mm.

The actual clear width of door required may be greater than this, depending on, for example, whether the direction of approach (from either side) is straight on, or at an oblique or right angle, or from a narrow route. The designer should also consider the extent to which the door may not be able to open to 90° allowing for the projection of door furniture, other furniture, or wall configuration. An effective clear width of less than 800 mm may result in damage to the door or to the person.

5.3 Circulation within the dwelling

5.3.1 Circulation within the entrance storey of the dwelling

The objective of this section is to provide ease of access around the entrance or principal storey of the dwelling, into habitable rooms and permit use of a WC facility, which may be within a bathroom. The objective is not to facilitate any occupant for fully independent living; a home may still have to have adaptations according to the occupant's specific needs. The provisions are to enable people of all mobility abilities to be able to visit, and to accommodate a good proportion of people's general mobility needs throughout life.

The design provisions of the Approved Document use the general requirements of a wheelchair user, because a wheelchair user's movements can most easily be obstructed by inadequate design. However the provisions will also help people with prams, pushchairs, and walking aids.

The requirements of M1 and M4 mean that reasonable access must be provided to habitable rooms and to a WC on the principal or entrance storey of the dwelling. Where it is not possible to make the main entrance accessible and an alternative is provided instead, the internal circulation routes around the principal storey should be accessible from the alternative entrance and this may need careful design. It may be for instance that a level entrance is made through patio doors in one of the habitable rooms, in which case the WC and the other habitable rooms must be accessible from the room which contains the patio doors.

Doors, corridors, and passages need to be wide enough to conveniently accommodate a person in a wheelchair allowing for manoeuvring past local obstructions such as radiators and other fixtures. Doors need to be wide enough both for head-on and right angled approach. A wider door or corridor is needed when the door is not approached head-on. Figure 5.2 shows the rules for corridor widths, doors into habitable rooms and a WC room. The Approved Document makes it clear that the unobstructed corridor widths are for those within the principal or entrance storey, but is not clear for door widths. It would of course be sensible to ensure that these minimums are adhered to throughout the property.

The minimum corridor width is 900 mm, and, where obstructed by radiators or other fixed fittings less than 2 m in length, should not be less than 750 mm wide. No such fittings should be placed opposite a door opening. The narrower the corridor, the wider the door needs to be.

The provisions apply equally to the entrance storey of a block of flats, and should be interpreted accordingly, so that door widths and corridors are sufficiently wide. There are issues for access in the provisions for fire safety and the requirements of Part B. There are likely to be fire doors into stairways and corridors giving access to flats. The self closers provided to these fire doors should not be so forceful as to restrict access and egress for those people with low upper body strength and manual dexterity issues, including children and older people. BS 8300 suggests a maximum of 20 N closing force in most situations.

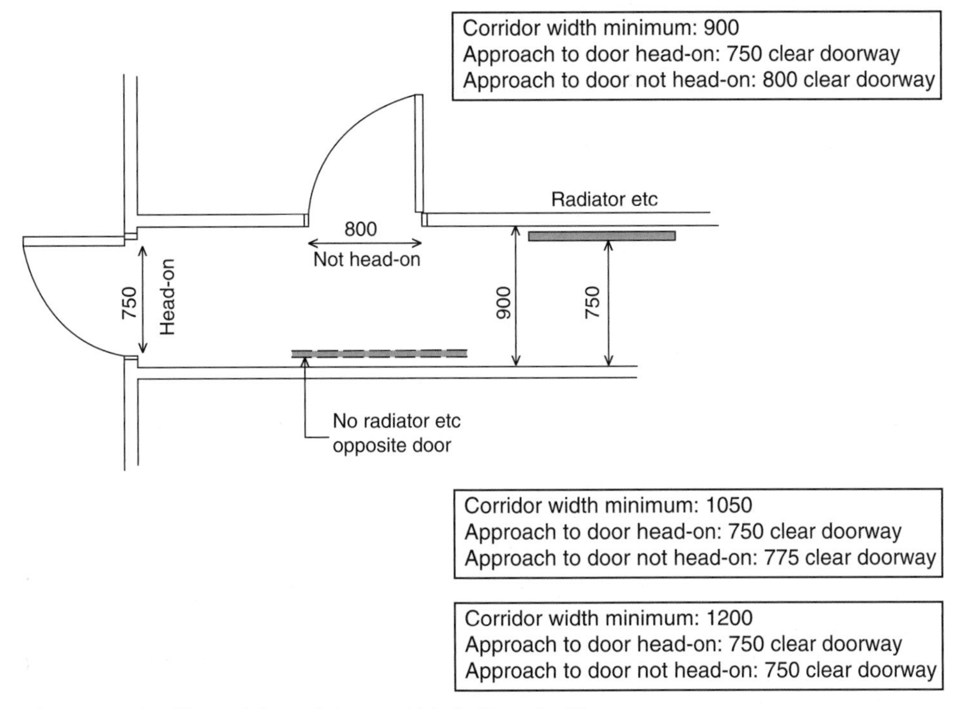

Figure 5.2 Corridor and door minimum widths inside a dwelling

5.3.2 Vertical circulation within the entrance storey of the dwelling

Steps within the entrance storey of a dwelling must be avoided if at all possible. Sometimes, in exceptional circumstances on a severely sloping plot, it may be not be possible to avoid putting a change of level involving steps, in the entrance storey. In this situation only a stair which is wide enough to be negotiated by an ambulant disabled person, with assistance if necessary should be provided. The stair must have handrails on both sides.

In these exceptional circumstances, any stair provided in the entrance storey which gives access to habitable rooms should have:

- A minimum clear width of 900 mm.
- Where the stair consists of three or more steps, a suitable and continuous handrail on both sides and on any intermediate landings.
- Dimensions of the risers and goings should be accordance with the guidance for a private stair as in Approved Document K (ADK). These are currently (ADK 1998 edition, amended 2000) set at :
  - Risers maximum 220 mm
  - Goings minimum 220 mm
  - Pitch not more than 42°.

5.4 Lifts and stairs in flats

5.4.1 Design objectives

Although it is acknowledged that people should be able to visit an occupant on any storey in a block of flats, and that a lift would be the most suitable means of vertical access for many people including those with mobility impairments, small children, and/or baggages; the Approved Document does not recommend that a lift is provided in all cases. This would seem to be surprising when lifts are required in public and commercial buildings.

It indicates that when a lift is provided it should meet certain standards and when a lift is not provided the stair should meet certain standards and states that Requirement M1 can be satisfied in either case.

5.4.2 Common stairs

Common stairs in a block of flats, where a suitable passenger lift is not provided should be designed to suit the needs of mobility impaired people and people with impaired vision. It should be designed with the following considerations:

- Step nosings which are clearly visible by the use of contrasting brightness.
- Top and bottom landings which follow the guidance for length given in ADK for Part K1. This is currently (ADK 1998, amended 2000) to be
  - level, no less wide than the width of the flight;
  - with a going at least as long as the flight width;
  - free from obstructions;
  - with any door swing across it such as to leave an area 400 mm wide across the full width of the landing.
- Uniform risers not exceeding 170 mm in height; (Note that ADK gives maximum recommendation of 190 mm for this type of stair but states "*For maximum rise for stairs providing the means of access for disabled people reference should be made to Approved Document M: Access and facilities for disabled people*". Of course Part M is not specifically for disabled people, but for all people, and as mobility impaired people and other people who find large risers problematic, are included in 'all people', then their requirements will need to be taken as the general requirement.
- Uniform goings not less than 250 mm in length.
- Tapered treads not less than 250 mm measured at a point 270 mm from the inside of the tread.
- Risers which are not open.

- Any nosings do not project more than 25 mm, so that they do not constitute a trip hazard. It is very easy for someone who cannot manipulate their feet and ankles readily, to catch the toe of their shoe under a projecting nosing, causing them to fall forward.
- A suitable and continuous handrail on both sides of the flight and landings if the rise of the stair is two or more risers.
- The handrails project at least 300 mm from the top and bottom riser.
- The handrail positioned 900 mm above the pitch line of the stair and 1000 mm above the top, bottom and intermediate landings.

Some of these design considerations are given in Figure 5.3.

5.4.3 Lifts

The Approved Document contains advice for accessible lifts and further information provided here is held in the British Standard (BSI, 2001).

Although they are the most convenient way for many people to reach the upper floors, it is not a stated requirement in the Regulations, nor given as recommendation to meet the Regulations in the Approved Document that passenger lifts are provided in blocks of flats. If a lift is provided then it is a requirement that it meets certain accessible standards. It should be usable by unaccompanied wheelchair users and people with sensory impairments. It should also contain suitable delay systems to enable people who require it, more time to enter and leave the lift car and lessen the risk of contact with the closing doors.

The internal space of the lift car should be of minimum dimensions to permit entry by most wheelchairs. The wheelchair user should be able to wheel themselves into the car, use the controls still facing forward, and then leave by reversing out. To do this the floor space of the car should be wide and long enough and the controls positioned within easy reach from a forward-facing sitting positioning. If the wall facing the doors contains a mirror, this helps the wheelchair user in identifying which floor they are on (albeit in 'mirror-image') and if anyone is in the way when reversing. However, a mirror is not a recommendation given in the Approved Document. There should be sufficient unobstructed space outside the lift doors for a wheelchair user to have turning space before or after they leave the lift car. The call control plates for the lift should also be accessible from a sitting position. Reasonable heights for the controls inside and outside the lift also result in ease of use by people of short stature, those with limited use of arms and by older children.

The controls should be easily found and used by people with sensory impairments. For example, people with impaired vision need to be able to easily call a lift, have suitable audible information telling them that the lift has arrived, have suitable tactile indication to help them select the right floor, and for a lift serving many floors let them know that they have arrived at the right floor.

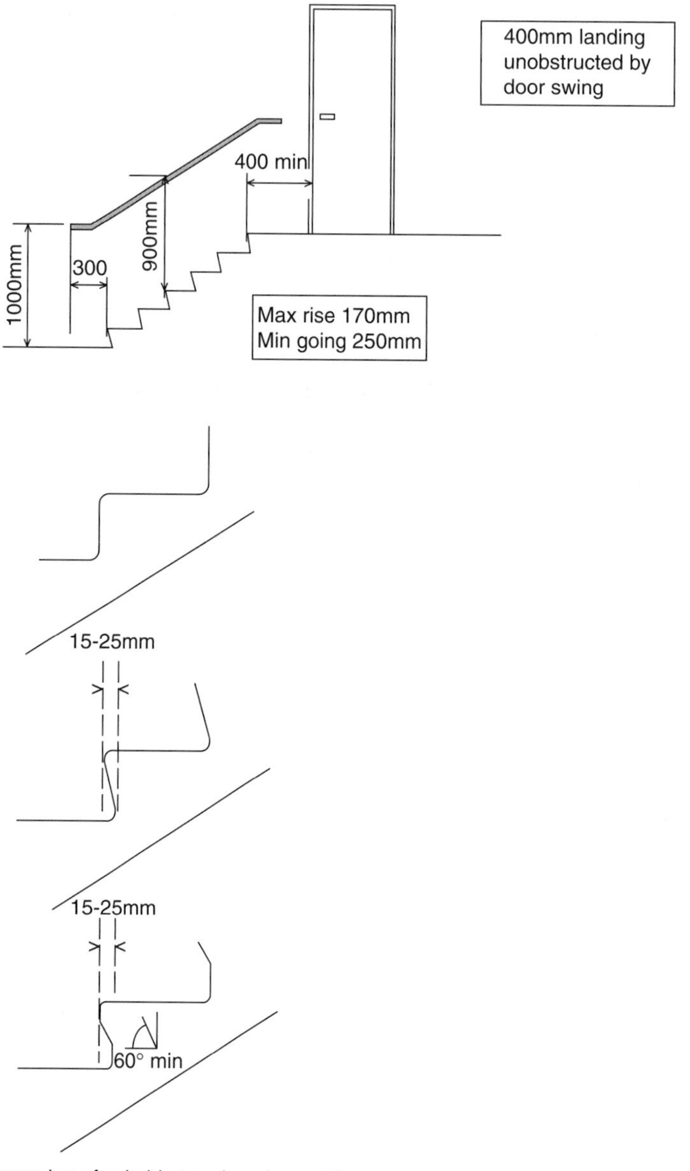

Figure 5.3 Approved Document K and M recommendations for common stairs to flats

The use of visually and acoustically reflective wall surfaces can cause discomfort for people with visual and hearing impairments. Excessive acoustic reverberations can affect people's abilities to distinguish speech and other sounds. It is a recommendation for lifts in buildings other than dwellings that such surfaces are not used, and therefore, similar measures should be considered for lifts in buildings containing dwellings. This is not a contradiction of the recommendation for a mirror, but care should be taken in the choice of mirror. The mirror should also not extend below 900 mm from the lift floor to avoid confusing people with impaired vision.

Colour contrast is useful to some people with impaired vision, so that the lift call buttons, the lift doors, and the lift car controls should be easily distinguishable in colour or luminance from the surrounding walls.

Other helpful features are that the lift floor should be slip-resistant, with similar frictional qualities to the landing floor to avoid the risk of stumbling, and also have a high luminance to reassure people with visual impairments that they are not stepping into an open lift shaft.

If a passenger lift is provided, to meet the requirements of M1, the Approved Document states that a suitable passenger lift would require the following features (see Figure 5.4):

- A minimum load capacity of 400 kg.
- An unobstructed, accessible landing space at least 1500 mm × 1500 mm in front of the lift doors.
- Have a door or doors which open to provide a minimum clear opening width of 800 mm.
- Have a car with minimum dimensions of
  - Width 900 mm
  - Length 1250 mm.

Although other dimensions can be used if demonstrated as being suitable for an unaccompanied wheelchair user.

- Have landing and car controls at the following positioning dimensions
  - Minimum 900 mm above the landing or lift car floor
  - Maximum 1200 mm above the landing or lift car floor
  - On the side wall of the lift, a minimum of 400 mm from the front wall.
- Have tactile indication showing what storey it is on each landing adjacent to the lift call button.
- Have tactile indication on or adjacent to the lift car control buttons to confirm the floor selected.
- Incorporate a signalling system which gives a visual warning that the lift is answering a landing call.
- Have a 'dwell time' of 5 s after the doors are fully open before they start to close. This system may be overridden by an electronic door re-activating device, but not a door-edge pressure system, provided that the minimum time for a lift door to remain fully open is 3 s.
- Have visual and audible indication of which floor has been reached, when the lift serves four or more storeys, including the ground floor.

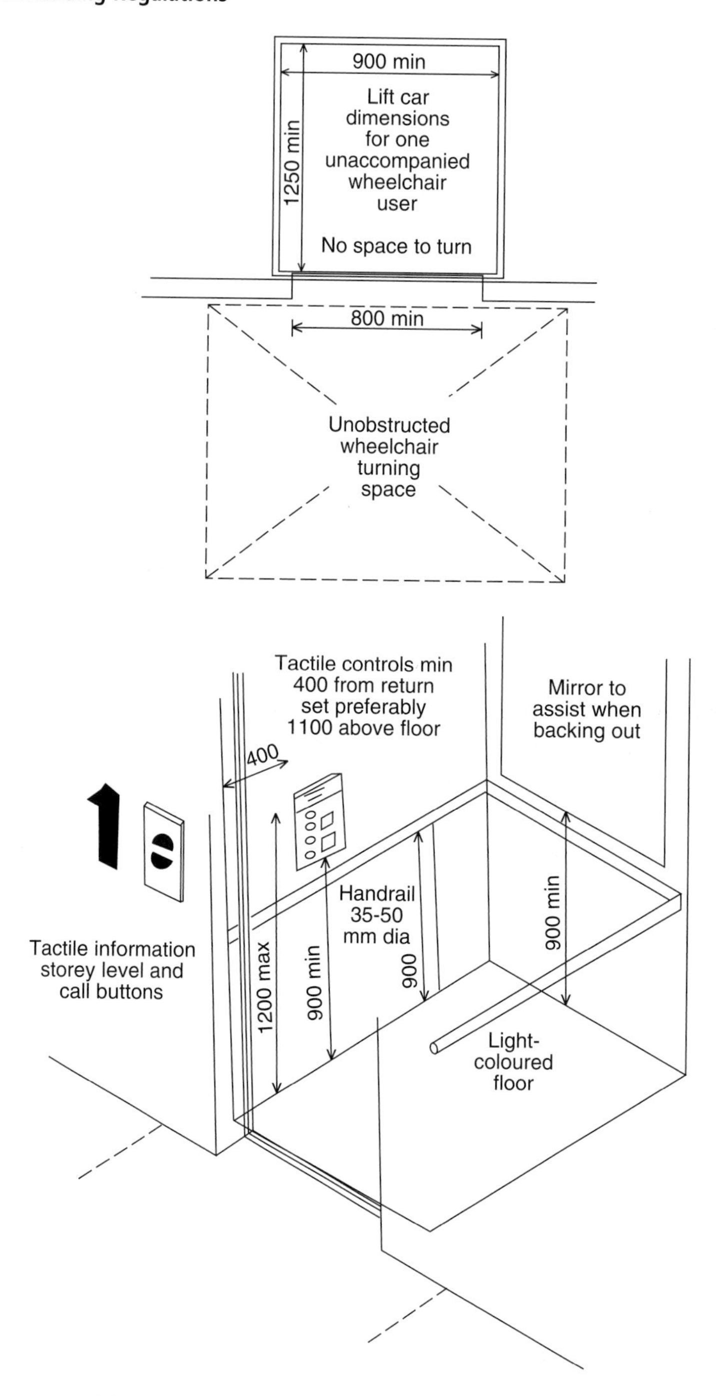

Figure 5.4 Passenger lift dimensions in residential buildings

5.5 Facilities in dwellings

5.5.1 Switches and sockets in dwellings

It is a requirement of M1 that reasonable provision is made for people to use the facilities in a building. To do this the person should be able to reasonably use the sockets and other outlets available. This section aims to facilitate easy use of switches and sockets in dwellings. The custom has been for sockets, particularly, to be placed generally at skirting level. There is no specific reason for this to be the case. Elderly people, for example, people with arthritis, mobility impaired people, and people who use wheelchairs often have more limited reach than others and find it more difficult to bend down to these sockets.

Therefore, switches and socket outlets for such things as electrical appliances, lighting, television aerials, telephone jack points, etc., should be mounted at suitable heights above floor level so that they can be more easily reached. Essentially this means locating sockets and switches in habitable rooms between 450 and 1200 mm above finished floor level.

The diagram accompanying this section, in the Approved Document suggests that this requirement should extend to door bells and entry phones.

It is helpful if these outlets, switches, and controls are positioned consistently within a dwelling in relation to the floor and doorways to ease location, for example, aligning light switches with door handles. Socket outlets whose plugs are frequently removed and replaced are better placed at the top of the height range. The higher the socket outlet the easier it is to push in or pull out the plug. It is also helpful if plates contrast with their surroundings. These suggestions are not a requirement of Building Regulations but can be found in the British Standard (BSI, 2001). Figure 5.5 illustrates recommended positioning for both ADM and BS8300.

5.5.2 WC provision in dwellings

The objective of this section is to ensure that there is an accessible WC in, or close to, the entrance storey of a dwelling, enabling visitors of all abilities to use the facilities. It is requirement of M4 that reasonable provision is made in the entrance storey of a dwelling for sanitary conveniences.

The provision of accessible facilities is also helpful to the occupant of a dwelling during various stages of their life, including their family's or dependant's lives. Again it is not designed to afford fully independent living for all people, and adaptations may need to be made to suit specific needs.

Where the entrance storey contains no habitable rooms it would be reasonable to expect that visitors would be entertained on a storey where there are habitable rooms and that these would be on a floor described as the "principal storey" being a storey nearest to the entrance storey but containing habitable rooms.

The Approved Document states that the WC should be located so that there is no need to negotiate a stair to reach the WC from the habitable rooms in the

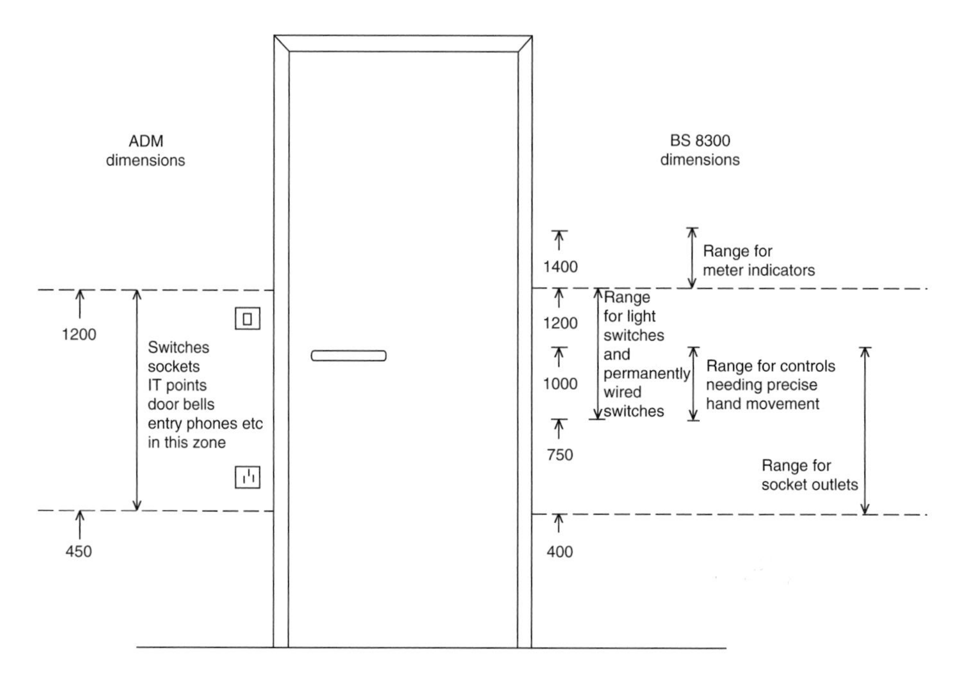

Figure 5.5 Positions of sockets, switches etc

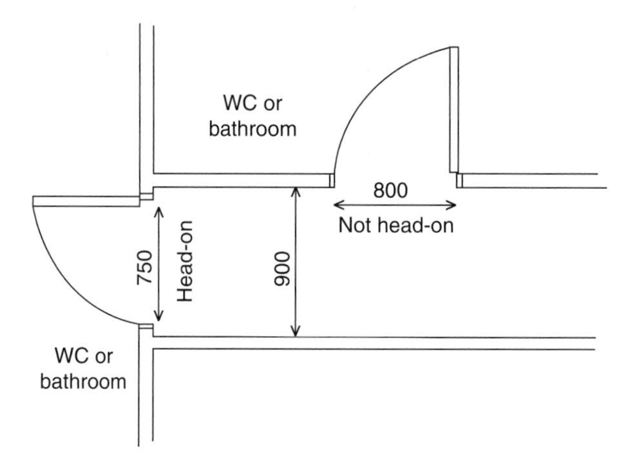

Figure 5.6 Residential door widths into WC

entrance or principal storey. It does not have to be a separate WC compartment but facilities could be located in a bathroom.

The door width requirement is the same as those required for internal circulation at Figure 5.2 used above, repeated at Figure 5.6 for ease. A wider door than the minimum stated would permit easier access and manoeuvrability for people in wheelchairs, especially when it is necessary for a person to enter the WC accommodation from an angle rather than head-on.

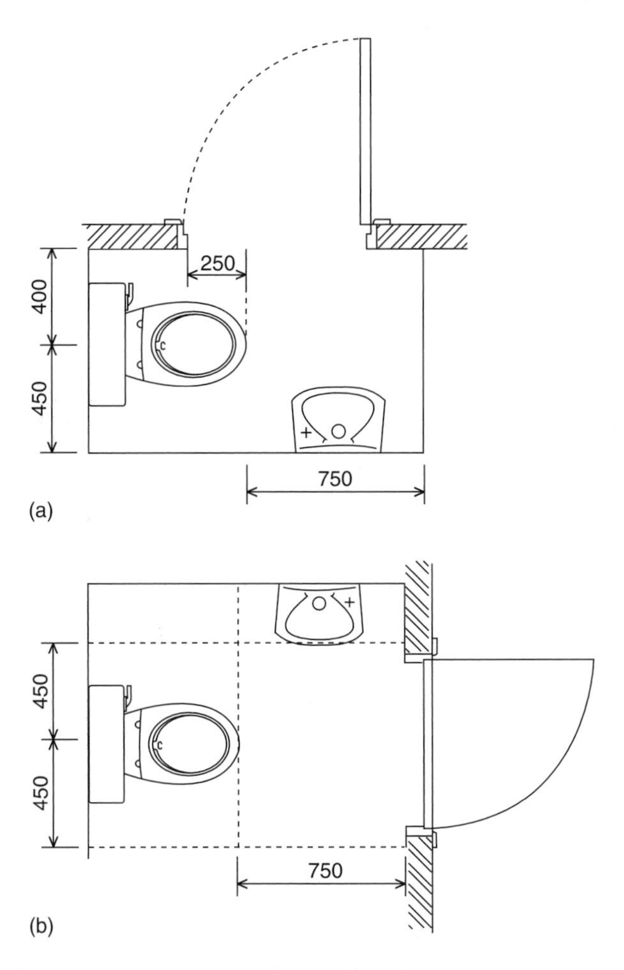

Figure 5.7 Typical WC compartments in accordance with ADM

Figure 5.7 shows designs of a WC compartment in the entrance storey of a dwelling in accordance with the recommendations given in ADM. The Approved Document does not suggest a position for the wash basin, restricting itself to stating that the basin must not impede access. A WC compartment should include a wash basin and this is of most use if usable while seated on the WC. The ADM requires a clear space of 750 mm long, suggested as 1000 mm wide (or at the very least 900 mm wide) in front of the WC, but does not comment whether the basin can be in this area. If the basin is wall mounted, of minimum size and close to the WC then the additional requirement that a wheelchair be able to approach within 400 mm of the WC could be met.

The Approved Document states that it may not always be practical for a wheelchair to be fully accommodated inside the WC compartment. This means that a wheelchair user can use the WC but the door may have to be left open to accommodate the chair.

If the dimensions are used that are given in the two illustrative sketches in the Approved Document at Diagrams 31 and 32, the door will have to be open, and the corridor blocked in both cases. This reduces the dignity of any wheelchair user and should be avoided. It is likely that the Approved Document has in mind the scenario where a wheelchair user is able to leave their chair unaided, transfer to the WC, close the door while seated on the WC, and retrieve their chair afterwards. The designer should appreciate this and design appropriately.

Diagram 32 of the Approved Document is unclear, particularly in relation to the dimension of 250 mm, whether this is minimum or maximum, its significance, and in relation to the position of the edge of the door frame. Figures 5.7 and 5.8 give an interpretation of these ambiguities.

The recommendations are that the wheelchair can get to within 400 mm of the front of the WC before being impeded by the door frame and walls etc, to enable the person to transfer between chair and WC (see Figure 5.8).

Considering the transfer from chair to WC and back, it is likely that the person will need to use the toilet seat to be able to steady or lift themselves. It is therefore important that there is a washbasin positioned both easily accessible from the wheelchair position, and not impeding the access of the wheelchair as far into the WC accommodation as possible. The person will possibly want to wash their hands both while seated on the WC and again when transferred back into their wheelchair. This is not a stated recommendation in the Approved Document, which does not mention basin position for residential property, but it is a very practical consideration that the designer should take account of.

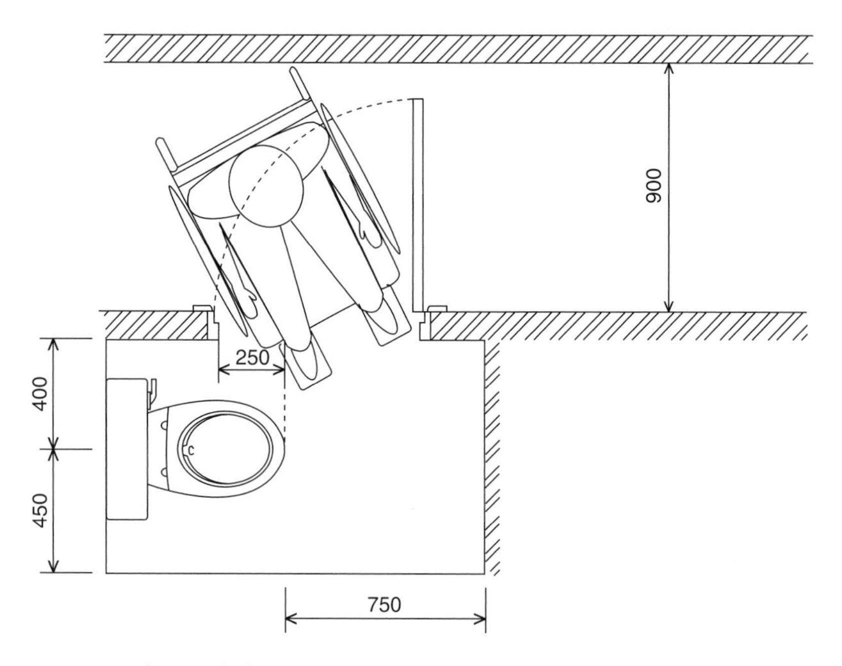

Figure 5.8 Position for transfer between wheelchair and WC in accordance with Approved Document M diagram 32 minimum standards

The designer should also consider the fixings of the basin and whether it is likely to be used for supporting a person seating themselves on or lifting themselves from the WC, and therefore how robust those fixings should be.

The following provisions apply to the design and location of the sanitary accommodation in the entrance or principal storey of a dwelling:

- A suitable and accessible WC is required in the entrance storey of the dwelling.
- Where there are no habitable rooms in the entrance storey, a suitable and accessible WC is required, located in either the entrance storey or the principal storey (i.e. the one containing a habitable room) of the dwelling.
- The access route, corridor, etc. serving the WC accommodation from the habitable area should have an unobstructed width in accordance with Figure 5.6.
- The door of the WC accommodation is to open outwards.
- The clear opening width of the door should be at least as wide as those shown in Figure 5.6 and preferably wider.
- The door of the WC accommodation should be positioned so that a wheelchair user can access the WC. A wheelchair should be able to approach within 400 mm of the WC.
- The WC compartment should provide an clear unobstructed space for a user as shown in Figure 5.7.
- The washbasin should not obstruct the clear space, even if wall-hung. The basin may be used for support and should be fixed accordingly.

Typical WC compartments in accordance with ADM are illustrated in Figure 5.7.

6

BS 8300 and The Disability Discrimination Act

6.1 Introduction

The Approved Document M (ADM) is not the only available guidance to Part M of the Building Regulations. Although it may usually be the most suitable guidance, there may be occasions when the advice is not reasonable, appropriate or most detailed. Other suitable guidance can be used in place of ADM. This should be explained in the accompanying access statement to be provided with the Building Regulation application. One such guidance document is BS 8300. This document also gives valuable advice when any physical adjustments need to be made to buildings so that the service provided in the building is not discriminatory to disabled people.

The chapter looks at BS 8300, comparing it to ADM and discusses some of the requirements for the *Disability Discrimination Act* (DDA) in comparison to those of the Building Regulations Part M.

6.2 BS 8300

British Standard, BS 8300:2001, is the Code of Practice for the design of buildings and their approaches to meet the needs of disabled people (BSI, 2001). It is based on ergonomic design and offers good practice guidance. In its Foreward, it sets out its purpose: it provides guidance on good practice in the design of domestic and non-domestic buildings and their approaches so that they are convenient to use by disabled people. The recommendations relate to general or common elements of buildings and their facilities, and also to specific building types. It is very useful in covering initial design information needed which may not be subject to Building Regulations. For example it gives guidance on the appropriate numbers and proportions of accessible car-parking spaces for different uses.

The Standard identifies its shortcomings with regard to requirements for people with sensory impairments, for example those who have hearing or visual disabilities. It states that at the time of its production, further research was needed into the risks and inconvenience presented by buildings for many such people. It also presents other areas of research ongoing or required. This is useful knowledge for the designer as it makes them aware that current guidance is simply that – current at the time of publication. The designer can then, if relevant to the work in question, carry out further investigations to determine if other guidance, perhaps published later, gives additional useful information.

It is a comprehensive and detailed document, providing guidance in some areas not covered by building regulations, which can be very useful to designers interested in designing for the usability and viability of the building. Once the building is occupied, other areas of legislation are triggered, including the DDA, and the designer requires guidance for this. The British Standards offers much of interest and usefulness.

The document suffers from a lack of an index and it is hoped that this will be provided in future editions. Good scrutiny of the Contents pages, particularly the Figures and Tables, can often help in finding the clauses needed.

At publication, this Standard was a landmark document and provided the most current advice available with much research and information pulled together in one document. A large proportion of the content has been used or built upon by the ADM, 2004 edition.

The Standard gives a lot of background information and explanations needed to understand the reasons for recommendations. It is important to understand these reasons behind the recommendations so that where the circumstances give a choice, for example induction loop or infrared hearing enhancement, the most appropriate choice can be made. In other situations, it may be that the full recommendations may not be reasonably possible, and background reasons and explanations allow the designer to make the most appropriate and helpful compromise.

BS8300 is a different document with a different purpose to ADM; they have therefore differences as well as conformity with each other. Included at Appendix 1 of this book is a chart of some differing interpretations.

BS8300 is offered as suitable guidance for design throughout ADM, for example the note at clause 4.35 of ADM states:

> '**Note:** Detailed guidance on surface finishes, visual, audible and tactile signs, as well as the characteristics and appropriate choice and use of hearing enhancement systems, is available in BS8300'

Use of ADM is only one way of complying with Regulation M of the Building Regulations; it is only guidance and does not have to be complied with. Although in many cases ADM provides the standard, the benchmark for compliance, in some circumstances other guidance would be more appropriate, more suitable, more up-to-date. The British Standard is suitable guidance to use when complying with the Regulations. If there are specific circumstances where the British Standard gives direct and contradictory advice to that of the Approved Document, then the designer may wish to comply with Regulation M by using that advice given by ADM. However there must be a health warning codicil to that: if the British Standard's advice gives an enhanced level of access then this may be regarded as more reasonable by the courts following occupation, if the occupier is subject to the DDA.

6.3 Disability Discrimination Act 1995

6.3.1 General

The DDA 1995 (DDA) has two main parts – Part 2, Employment, and Part 3, Provision of Goods and Services. The parts have been implemented in stages.

In 1996, it became unlawful to discriminate against disabled job applicants and employees. The duty to make reasonable adjustments and to introduce auxiliary aids and services became compulsory. In 2004, these provisions, which applied to employers with 15 or more employees, were extended to all employees.

Also in 1996, for the provisions of goods, facilities, and services, it became unlawful to refuse a service to someone on the grounds of disability, to provide a service on different terms, or to offer a service of a different standard.

In 1999, the duty to make reasonable adjustments came into effect and those duties to provide auxiliary aids and services, and to provide a service by an alternative method where physical barriers prevent access to a service.

In 2004, the section of Part 3 came into effect such that service providers may have to make physical adjustments to their buildings and grounds so that the features do not prevent disabled people from using their services. The provisions of Part 3 of the DDA extended to Higher and Further Education.

Part 4 of the DDA covers education and this has a similar but different timetable.

Note that it is not possible for a building to 'be in compliance with' or 'comply with' the DDA. There are various reasons for this, the main one being that it is not the building which is required to not discriminate; that is the duty of the service provider, employer etc. In the case of Part 3, it is the service provided which should not be discriminatory, not any specific building. Any suggestion that a building complies with the Act is a misunderstanding of the Act.

Notwithstanding this, a building can of course be so designed or used that an obstacle is created which may result in a disabled person being discriminated against, but rectifying this would still not make the building to be in compliance, because it is the service which matters.

A service provider could be providing a service out of a building which has very poor access. The service provider may be able to alter their service so that it is not discriminatory, without altering the building. A typical example would be a hairdresser whose premises are on the first floor of a rented building. They may be able to provide a home visit service for those people who cannot mount the stairs. The service offered (that of hairdressing) is no longer discriminatory, but the inaccessible building has not needed to be altered.

The DDA is concerned primarily with people not property.

6.3.2 Definition of disability

The DDA is concerned with discrimination of disabled people and therefore a definition of disabled people is required. This is very different to Regulation M of the Building Regulations which is concerned with reasonable provision for people to access and use a building and its facilities. There is no definition of 'people' for Regulation M, so this phrase includes all people from birth to grave with whatever idiosyncrasies they may have due to age, infirmity, genes, syndromes etc. But for the DDA a definition is needed.

The DDA takes a primarily medical approach to disability and defines disability in Section 1.(1):

> Subject to the provisions of Schedule 1, a person has a disability for the purposes of this Act if he has a physical or mental impairment which has a substantial and long-term adverse effect on his ability to carry out normal day-to-day activities.

There is no exclusive definition of what disabilities are included in the DDA. The definition excludes short-termly impaired people such as those who are pregnant or who have a broken leg. The reference to 'normal day-to-day activities' means that people who cannot carry out activities which are not normal to most people, such as climbing Mount Everest, are not necessarily able to be defined as disabled. Which is not to say that a particular person who is disabled is not able to climb this mountain. The first disabled person to attempt Everest was American Tom Whittaker, who climbed with a prosthetic leg to 24 000 ft in 1989, 28 000 ft in 1995, and finally reached the summit in 1998 (Peakware, 2004).

Generally, when considering access audits to be used for DDA considerations the following ranges of disabilities should be taken into account:

- People who use wheelchairs
- People who have mobility problems
- People who have manual dexterity problems

- People who have a visual disorder
- People who have a hearing impairment
- People who have speech disorders
- People who have continence problems
- People who have mental disorders.

However there are many more impairments which may be encountered or experienced. The person discriminated against does not have to be registered as disabled.

6.3.3 Disabled access – employees

Under Part 2 of the DDA 1995, employers must not discriminate against employees or prospective employees because of their disability. An employee who becomes disabled is legally entitled to the same consideration.

Until 2004, any employer with 15 or more employees was covered by the employment duties in the DDA; this 15 persons threshold was eliminated in 2004 so that all employers are now covered by the requirements.

'Access to Work' provides practical support to disabled people entering or in paid employment to help overcome work-related obstacles caused by disability. The local Jobcentre is the contact for information for a specific situation. Therefore if a firm wishes to employ a person with a particular disability, or if an existing employee becomes disabled, and there are workplace problems or obstacles which would make it difficult for that person to easily do their job, 'Access to Work' may be able to provide help or grants to overcome these.

One of the duties imposed by the DDA is to make reasonable adjustments when any physical feature of the workplace or any arrangements substantially disadvantage a disabled person compared to a non-disabled person. These requirements are, of necessity, person-specific. They have to be tailored to the employee's specific requirements. For example, an employee who is profoundly deaf is undoubtedly "disabled" but may not need an accessible WC, however they may require a device to inform them when the fire alarm is activated. Or an employee may become disabled with relapsing-remitting multiple sclerosis and need their office moving to the ground floor for relapse periods when they have to use a wheelchair and may also need other aids, perhaps relating to eyesight or dexterity.

If alterations need to be made to the building, for a specific employee, these adjustments still have to comply with Building Regulations, including Part M.

6.3.4 Disabled access – service providers

Under Part 3 of the DDA all service providers must ensure that from 1 October 2004 any physical barriers, which may result in discrimination of disabled people who wish to use the service, should have been removed or overcome. These provisions affect all buildings, new or existing, which are accessed by the public or users of the service. Procedures and policies that may result in the discrimination of disabled people should have been replaced or overcome from October 1999.

Although the requirements are that works should have been done by October 2004, not all premises come under the Act because it is the service which must not be discriminatory, not the building. If the building provides an obstacle it could be altered, but the service may be provided in such a way that no discrimination is encountered whatever the condition of the building or buildings used by the service provider. In such a case there is no need to alter the building. Other service providers will of necessity need an accessible building with accessible facilities. It all depends on the circumstances and the way the service is provided.

Therefore as service providers move premises, the building may have to be altered. This is likely to be the case in existing properties, but newly built properties may also require alterations, again depending on the particular circumstances.

The Act is far-reaching and there is no exclusive definition of disability, apart from the definition limitations of long term and affect on normal day-to-day activities. However, the DDA is about ensuring that there is no unreasonable discrimination of disabled people in the areas covered. This is an essential difference from the Building Regulation Part M which is about access for people. Table 6.1 summarizes the comparison between Part M and the DDA.

There has been some consideration regarding buildings being constructed in compliance with Part M, becoming occupied, and then falling foul of a discrimination risk or claim under the DDA. Buildings themselves (not the service) have therefore been given a 10 year reprieve. The 10-year rule covers the following circumstance:

- The building is built in compliance with Part M
- A particular part of the building complies with Part M and the recommendations in ADM
- That part of the building causes an obstruction which results in discrimination of a disabled person under the DDA
- The occupiers cannot then be required to alter that part of the building under the DDA
- But the service does not get a reprieve and the service itself is likely to have to be altered so that it is not discriminatory.

Table 6.1 Short comparison of Part M and the DDA

	Part M	**DDA Part 3**
Factor(s) to be considered:	Building and facilities	Service provided
Sector covered:	People	Disabled people
Definition of disability:	No	Yes but not exclusive
Requirement:	Accessible building facility	Lack of discrimination
Can a building comply?	Yes	No
Does the building have to be accessible?	Yes if new or extended	No if the service can be offered in a different but acceptable way.

There has been confusion with people thinking that a building in compliance with Part M is therefore in compliance with the DDA for 10 years. That cannot be the case because the building cannot comply as it is the service which must not be discriminatory.

So a handrail that has been provided in accordance with Part M, which later is found to cause an obstruction to the public because of the way the service is provided, does not need to be changed for 10 years, but the service will need to be altered so that it is not discriminatory.

One cannot use the idea that if a building has been built in conformance with Part M it is somehow exempt from the requirements of the DDA.

6.3.5 *Duty of service providers to make adjustments*

Section 21 of the DDA clarifies what the duties of the service providers are in this respect. Section 21, subsection (2) states:

> Where a physical feature (for example, one arising from the design or construction of a building or the approach or access to premises) makes it impossible or unreasonably difficult for disabled persons to make use of a service, it is the duty of the provider of that service to take such steps as it is reasonable, in all the circumstances of the case, for him to have to take in order to -
>
> (a) remove the feature;
> (b) alter it so that it no longer has that effect;
> (c) provide a reasonable means of avoiding the feature; or
> (d) provide a reasonable alternative method of making the service in question available to disabled persons.

There is no level of priority given for these steps. It can be as acceptable to provide an alternative method of providing the service as to remove the feature. If the person does not feel discriminated against, there is no discrimination of them. However the next person may feel discriminated against and need different consideration. For this reason, one cannot say that a service is in compliance with the DDA. Similarly, a building cannot be in compliance with the DDA. It would all depend what service was being offered which involved the building, how the building was involved and what the particular disability of the person was.

There is also no requirement under the DDA to carry out any prescriptive measures to alter, amend or change the building. It may be that to ensure a particular service is offered in a way that is not discriminatory, the service provider may decide that a particular alteration or addition would be reasonable, but this may not be the case for another service provider. Certainly blanket measures for any building are likely to be at best unnecessary and at worst unhelpful.

6.4 Access Audits

6.4.1 General

An access audit is a survey conducted on a building specifically to consider the premises, in its current use under the auspices of the DDA. This necessarily involves looking at the working practices of the occupier because it is the way that the service is provided that may be discriminatory not the building itself, although the building may present obstacles to a person attempting to access or use that service.

A person conducting an access audit therefore has to be competent in the dealings of the DDA. The test is whether or not the obstacle makes the service impossible or unreasonably difficult to access or use, due a disability suffered by the person who is therefore discriminated against. If it is not impossible or unreasonably difficult, just a bit of a nuisance, then it may not be discrimination. The person conducting the access audit therefore has to be conversant with the problems people with disabilities encounter due to their disability and due to the environment, and be able to suggest appropriate and suitable alternatives, either to the way the service is offered or to get around the particular obstacle.

These are not skills that are acquired without specific, special and careful study and experience. Limited competency often results in inappropriate or unnecessary work. Access consultants and auditors registered with the National Register of Access Consultants are recognized to have achieved a measurable and tested level of competence in these matters.

As a minimum an Access Audit should identify the areas and specific points that need to be considered for different types of disability with the following including for each issue:

- **Physical feature** – the obstacle or thing in question clearly identified.

- **Acceptable features** – these should be described and explained, with reference to specific guidance, how the feature is acceptable. This is important for maintenance and repair programmes for clients to know what is good and acceptable about their properties and which should not be altered adversely.

- **Problems** – problems should be clearly identified, with reference to specific guidance and the type of person or nature of disability likely to be adversely affected by the physical feature. Without the specific guidance referenced the information is questionable as readers cannot judge on what basis the comment is being made.

- **Comments** – recommendations and notes as to how the problem might be overcome so as to avoid discrimination. As with any problem there will be more than one answer. Each problem may also have some consecutive solutions with different priorities. For example a particular obstacle may make the service impossible to access, a 'full' solution may be too expensive, but an immediate step could be taken costing little if anything which would improve matters, this would be priority 1, then a planned programme of improvements given priority 2 could be put in motion. Solutions which only amend the building without considering the procedures and practices of the way the service is offered are probably not helpful or effective.

- **Priorities** – this should give a professional judgement to provide a prioritisation for the recommendations.

Table 6.2 is a commonly used form of prioritization although there are other forms. The explanations are given in tabular form for clarity.

Table 6.2 Priorities scheme for Access Audit items

Priority	Explanation	Example
1	Access or use of service impossible or unreasonably difficult, therefore contrary to DDA. Highest priority	WCs provided for public use but no accessible WC
2	Access or use of service difficult, alter as soon as possible. Good practice	Unhelpful signs
3M	Alter as part of normal maintenance works i.e. when maintenance is required then change for a better solution	Decorative colours do not give good contrast
3R	Alter as part of refurbishment i.e. when refurbishment is required then change for a better solution	The handrail to the stairs is not of the recommended profile
4	Alter or improve when a specific need is identified e.g. when an employee needs it changing	The staff entrance has a short flight of stairs
X	No action reasonably practical	Small, 4-storey listed building with late 19th century lift which is not designed as recommendations and cannot accommodate standard wheelchairs *Note*: This would be for this specific point – many other adjustments are likely to be feasible in other areas. For example good signage to the lift, Braille buttons for calling the lift.

Costings or cost bands may be provided if requested. The BCIS publication, Access Audit Price Guide (BCIS, 2002) can be helpful here.

6.4.2 Scope of an audit

The scope of the audit should be to consider the status of the building with regards to the DDA and to suggest priority areas for action. The British Standard 8300:2001 'Design of buildings and their approaches to meet the needs of disabled people – Code of practice' is a good reference document in relation to the physical form of the building and its approaches.

It should be noted that while the British Standard gives details of best practice and good design, the DDA merely requires that the features of the building do not 'make it unreasonably difficult or impossible for disabled people to make use of their premises'. The report recommendations are therefore prioritised in respect of the requirements and limitations of the Act. However, as best practice, further consideration should be given to any other issues if subsequently raised by individuals.

Where recommendations are raised which would require physical adjustments, approvals under various legislation or policies may be required. Where physical adjustments cannot be made, management systems that can provide a solution should be put in place.

6.4.3 Formulation of an access action plan

Following the Access Audit, an Access Action Plan should be prepared and implemented. This may include reference to a Planned Maintenance Programme, decoration schedule, major/capital works budget and so on.

The DDA requires that 'reasonable' alterations are made to the building to remove barriers to access. The Disability Rights Commission give guidance in their Code of Practice as to the factors that could be taken into account in determining reasonableness if their impact was disproportionate to the benefit.

The law requires that these changes are reasonable, and when considering what would be reasonable, factors such as the size and the resources of the service provider, the extent that the physical feature is restricting access to services and the amount the service provider has already done will all be taken into account. So what is expected of the large retailer, for instance, will be very different to what is expected of a village hall.

The Disability Rights Commission recommends that service providers take a strategic approach to these requirements.

For larger service providers, where there are several premises and, may be even a head office, then an organizational wide approach should be taken to the planning of changes to ensure efficiency and consistency. The main thing for all service providers is that changes are planned, prioritized and coordinated.

Sometimes there may be a management solution to improving access to a service. Sometimes it may be more useful to make wide ranging changes to a small number of buildings rather than piecemeal changes to all buildings.

These are decisions which need to be taken by management and which require further long term thought to ensure that access for disabled people is not something special but something integral to the way that services are delivered.

10 Point plan

Some businesses are adopting a 10 point plan to fulfil their duties under the DDA. This plan enables them to appreciate where, when and how their services affect people with disabilities, and therefore to avoid discrimination. The following is based on a plan devised by the Navigator Group Ltd.:

(1) Understand the services offered
Seems obvious, but business need to understand their offering in the light of the DDA. The DDA is not essentially about buildings; it is about jobs and services. The building can be only a part of the challenges to disabled people who are trying to get a job, stay employed, or access a service. It is therefore important to understand, for Part 3 of the DDA, what services are being offered, how they are being offered, and the full scope of these services.

(2) Take a less abled perspective
For this, one needs to consider a whole range of disabilities individually, and even so some disabilities may be neglected. General advice recommends visual disorders, hearing disorders, mobility impairments, wheelchair users, mental impairments, manual dexterity disorders, continence problems, and upper body strength impairments are all considered. Remember that some people's requirements may be contradictory to others' and therefore there may need to be a range of different measures. A classic example of this is that wheelchair users prefer a shallow ramp, a person with foot or ankle problems prefer a step, and most people would like it to be level.

(3) Solution not excuse
Making a service universally accessible is not always easy. There may be conflicts. Ideal works may be costly. There may be other challenges such as listed building status, fire safety, health and safety. However, it is rare that satisfactory compromises cannot be achieved. Do not use these challenges as an excuse. Do the easy stuff first, meeting needs when there is no challenge, only thought required, and prioritize the rest. As shown above, accepted priorities are:

I. Health and safety – but this is really not to do with DDA and should be achieved anyway

II. Issues which make access to the service impossible or unreasonably difficult
III. Issues which make access difficult or awkward or annoying.

Priority III can be further subdivided into measures which can be easily into schedules for repair, maintenance, redecoration, replacement or refurbishing. If changes are built into these planned activities then additional costs are minimized or even negated. Changing the decorations at next redecoration for a helpfully contrasting scheme costs no more than a business would spend anyway.

(4) Build-in Access
This is a continuation of the last point. If the business is altering, amending, extending changing, then do so with improvement to access for the whole service in mind (buildings and site included).

(5) Information
It is not the building, it is the service. How do people know about the service offered? How do people get to the premises, point of access, etc? Make sure that any information provided can be received by all existing, new and potential customers or clients. This would include signage, TV/radio advertisements, leaflets, brochures, website, Yellow pages, other directories, publication advertisements, guide books, instruction books, receptionists, telephonists, letters and so on.

(6) Equipment
If particular or special equipment is needed, this should be accessible to and usable by people with differing needs or problems.

(7) Staff Training
This is essential, staff are front line. If a person does not feel discriminated against, there is no discrimination. Staff can alter the way services are provided or delivered, can see potential problems, can meet potential problems, all to the benefit of customers and clients as well as to the business. However to do this for the DDA, training in disability awareness is usually vital.

(8) Stay Accessible
Ensuring accessibility is not a one-off exercise, it is ongoing. Staff change, buildings change, services change, methods change. Constant review and awareness are needed.

(9) Ignorance is not a virtue
If you do not know, ask. Advice can be sought from the DRC, professional bodies, disability groups (public-spirited disabled people who can advise and recommend) and customers themselves. The customer group probably have a whole range of major or minor ability problems and can be enlightening.

(10) Do as you would be done by
This takes us back to rule 2. Put yourself in the customer or client's position and what would you expect?

6.4.4 Means of escape in case of fire

Safety in case of fire is an extremely important issue in buildings.

There are aspects of the traditional approach to fire safety that cause difficulties for some members of society.

Audible fire alarms may not be recognized by people with severe hearing impairments and escape signage may not be obvious to people with sight impairments.

People with mobility impairments including wheelchair users may not be able to use escape stairs.

However, the assumption that allowing access would automatically contradict fire safety responsibilities is not acceptable, there are many reasonable solutions to the problems highlighted above.

Escape issues are different to the matter of convenient access. A wheelchair user who has difficulties standing for long periods or walking more than a particular distance may well be able to use a traditional escape. In a well-populated environment people with hearing impairments may pick up on others reaction to fire alarms.

Safe means of escape can be dealt with in a number of ways, for example:

- Adaptations can be made to the building such as flashing beacons fitted to the alarm system to aid recognition for those with hearing impairments.

- People with specific hearing or sight problems can be issued with a vibrating device or other equipment to specifically alert the carrier to a danger or evacuation warning.

- Lifts are an important, sometimes essential feature of tall buildings for the evacuation of many ambulant disabled people and wheelchair users. Lifts CAN be used in a fire situation, but they need to be of evacuation standard, or a fire fighters lift. These are described in BS 5588-5 and 5588-8. An evacuation lift has a more protecting structure than an ordinary lift. In addition, it gets around the problems of lifts in a fire situation.

  One of the main issues is that people in a lift trying to make their escape from upper storeys may stop at the fire floor and the doors open on to the fire. An evacuation lift picks a person up from the designated floor and then returns to the ground floor or final exit floor to discharge safely. Its operation can be overridden by firemen so that it will not discharge its occupants into danger. It can then be instructed to rise in the building again, stop at a specific floor to pick up, for instance, a person in a wheelchair and return again to the final exit floor without stopping at any other intermediate floors.

An evacuation lift has its drawbacks, in expense and in slowness in picking up larger numbers of people on separate floors in a fire situation, but the advantages can far outweigh these. In a non-fire situation it acts as a normal lift.

A fire fighting lift is again a lift that is designed to have additional protection against fire and includes controls to be used by the fire service in fighting a fire. It is usually larger than a standard lift as it it also design to carry fire service personnel with all their breathing apparatus and equipment to the fire floor.

- Fire protected refuge areas can be identified for persons unable to use stairways in order that they can safely await rescue by suitably trained persons. Such an area would need communication equipment easily accessible for the disabled person to use, linking them to the persons who will effect their rescue. Notwithstanding this, refuges should be used with care and as a last resort. Putting yourself in the position of the person left in the refuge, while others all around are making their escape, with only the hope of someone rescuing you in time, gives an indication of refuges' undesirable features.

  Designers need to consider how many wheelchair users may in the building, or on any particular floor, and cater for that number accordingly. It would not make sense to provide for example for 5 wheelchair spaces in an auditorium, and then provide a refuge for only one.

  Refuges are possibly not acceptable for a fire risk assessment as the fire authority state that they are not under a duty to necessarily rescue people from refuges. People should be able to escape under their own unimpeded efforts.

- Equipment such as evacuation chairs can be provided to enable assisted use of stairs. Such equipment usually requires training and nominated persons to operate or use it. Not all people who use wheelchairs are able to use such equipment even with trained assistance. People who cannot leave their wheelchairs and be adequately supported by any device such as this must be considered. In addition, with the expectation of someone helping the disabled person, adequate training, including manual handling must be undertaken. Lastly, there is the question of what happens when the trained person is not in the building or at the floor required at the time of the fire.

- Management procedures, for example nominated fire wardens, can be adopted to overcome physical issues. However, specific measures must be considered fully.

- In the case of staff, personal emergency evacuation plans can be prepared by the individuals and competent people to address issues and plan how escape will be made. This could identify any equipment needs e.g. vibrating pager linked to the fire alarm for somebody with a hearing impairment.

- Visitors could be made aware of the availability of vibrating or other devices, the position of refuges, and/or the availability of other equipment or action needed by way.

Where the person and their abilities are known to a building's management, for example an employee, and a personal emergency evacuation plan (PEEP) is prepared for that person, this can be extremely effective. An employee in the World Trade Centre did not escape in an earlier evacuation for a number of hours. Without changing his floor of work, a PEEP was prepared for him, and he was one of those who did manage to escape the building.

The issue of safe means of escape for all should be discussed with all parties including the relevant Fire Safety enforcement body. Further reference on means of escape for people with disabilities can be found in Approved Document B of the Building Regulations and BS 5588 part 8.

6.4.5 Access policies

An access policy should arise from the audit, action plan, company policies and procedures and is an informative statement for customers, clients and other users. It will individual to the service provider. It should set out the limitations as well as the accessible facilities of the service provider. However, blanket statements such as "We are sorry but we have currently no facilities for disabled" are unacceptable as well as most likely being inaccurate. What the writer probably means is that "we have no accessible wcs suitable for wheelchair use", or that "we are on the first floor and there is no lift".

For example, the first phrase was included in an advertising brochure of a small business with a visitor centre, and in this case they meant both the second and the third phrases. Shortly after receiving their access report the advertising brochure was changed to state:

> "Our visitor centre is on the first floor, accessed by a stairway. We do not have a lift or fully accessible sanitary accommodation. Please telephone……..for our access policy and to discuss your individual needs."

Their access policy was amended to the following:

> "It is our policy to help all visitors to …. to gain as much enjoyment from your visit as possible.
>
> Staff are trained to be aware of the different difficulties you may experience and give help as appropriate. Please ask.
>
> It is however with regret that we confirm that there is no level or lift access for wheelchairs users to our regular …. Tours, nor to our Clubroom. However please contact ….. on ….. if you are a wheelchair user, and we will endeavour to accommodate you and your guests on one of our Open Access Visits.
>
> For food hygiene reasons, unfortunately we are not able to permit guide dogs into all of our production areas, however we do have a waiting area for

dogs and can offer a visitor additional help if required. It would be most helpful to us if we were able to be advised in advance that such facilities are required.

We recommend that if you have other mobility or visual impairments and would like to visit, that you contact .... before visiting to ensure that there is a person ready to assist you in accordance with your needs.

We can provide written explanations of the tour if required. These can be printed in large type if requested beforehand.

Because of the noise of the process, it can be difficult for a visitor with hearing impairments to hear all of the tour, and we do not have a sound enhancement system. Please ask for our Written Tour.

We do have sanitary conveniences in our clubroom but they are not as yet of a standard associated with accessible wc's. Alterations are being planned.

We provide an experience that uses all the senses of sight, sound, smell, touch and taste to ensure that everyone enjoys their visit to the full. Please contact .... if you need further information, or if there is more help we can give you."

The policy and indeed access is not ideal, but the service providers have considered their options and limitations in greater detail, and have plans for the future, because they do not wish to miss out on the estimated 8 billion prospective 'disabled' visitors.

REFERENCES, BIBLIOGRAPHY and FURTHER READING

British Standard BS 5503-3:1990, *Vitreous china washdown WC pans with horizontal outlet. Specification for WC pans with horizontal outlet for use with 7.5 L maximum flush capacity cisterns.* BSI.

British Standard BS 5504-4:1990, *Wall hung WC pan, Specification for wall hung WC pans for use with 7.5 L maximum flush capacity cisterns.* BSI.

British Standard BS 5395-1, *Stairs, ladders and walkways — Part 1: Code of practice for the design, construction and maintenance of straight stairs and winders.* BSI.

British Standard BS 5588-5:1991 (1991) *Fire precautions in the design, construction and use of buildings – Code of Practice for firefighting stairs and lifts*. BSI.

British Standard BS 5588-8:1999 (1999) *Fire precautions in the design, construction and use of buildings – Code of Practice for means of escape for disabled people*. BSI.

British Standard BS 5839-1:2002 (2002) *Fire detection and alarm systems for buildings – Code of practice for system design, installation, commissioning and maintenance.*

British Standard BS 8233 *Code of practice for sound insulation and noise reduction for buildings.* BSI.

British Standard BS 8300:2001 (2001) *Design of buildings and their approaches to meet the needs of disabled people. Code of Practice*, BSI.

Building Cost Information Service (2002) *Access Audit Price Guide*. RICS Business Services Ltd., London.

Building Regulations 2000 Approved Document C – *Site preparation and resistance to contaminants and moisture* (2004) Office of the Deputy Prime Minister. The Stationery Office.

Building Regulations 2000 Approved Document K – *Protection from falling, collision and impact*. 1998 amended 2000 (2000) Office of the Deputy Prime Minister. The Stationery Office.

Building Regulations 2000 Approved Document M – *Access to and use of buildings* (2004) Office of the Deputy Prime Minister. The Stationery Office.

Centre for Accessible Environments CAE (1999) *Designing for accessibility*. CAE.

CIBSE (1994) *Code for interior lighting.* Chartered Institution of Building Services Engineers, London.

Department of National Heritage and the Scottish Office (1997) *Guide to safety at sports grounds.* The Stationery Office.

Department of Transport Local Government and Regions (DTLR) and Department for Environment, Food and Rural Affairs (DEFRA) (2001) *Limiting thermal bridging and air leakage: Robust construction details for dwellings and similar buildings.* The Stationery Office. Norwich.

Disability Rights Commission (DRC) (2002) *Code of practice rights of access, goods, facilities, services and premises.* The Stationery Office, www.drc-gb.org/law/codes.asp.

Disability Rights Commission (DRC) (2004) *Access statements: Achieving an inclusive environment by ensuring continuity throughout the planning, design and management of buildings and spaces.* Disability Rights Commission.

Drivers Jonas (2003) *Planning and access for disabled people – A good practice guide.* Office of the Deputy Prime Minister.

English Heritage (2004) *Easy access to historic buildings.* English heritage

Haddon, M. (2003) *The Curious Incident of the Dog in the Night-time.* Published by Jonathan Cape. The Random House Group Ltd., London.

Hanson, J. (2004) *The inclusive city: Delivering a more accessible urban environment through inclusive design.* University College London.

ICI Plc (1997) *A design guide for the use of colour and contrast to improve the built environment for visually impaired people.* RNIB/GDBA Joint Mobility Unit and ICI plc.

Lacey, A. (1999) *Designing for Accessibility – an Essential Guide for Public Buildings.* Centre for Accessible Environments (CAE). London.

Navigator Group, *Ten Point Plan,* Navigator Group Limited, Kent House, Romney Place, Maidstone, Kent ME15 6LH.

Office of National Statistics (2002) *Living in Britain: Results from the 2001 General Household Survey.* Norwich, HMSO.

Peakware World Mountain Encyclopaedia (2004) Available from: http://www.peakware.com/encyclopedia/peaks/everest.htm. [Accessed 29 November 2004]

Planning Policy Guidance 15 (PPG15) *Planning and the historic Environment.* ODPM.

Royal National Institute for the Blind (RNIB) (1995) *Building sight: A handbook of building and interior design solutions to include the needs of visually impaired people.* HMSO in association with RNIB. London.

RNIB *See it right – Getting your message across.* RNIB

Royal National Institute for Deaf People (RNID) (2002) *Induction loop and infrared systems – information for people managing public venues.* RNID.

The Building Regulations 2000 for England and Wales (SI 2000/2531) (2000) The Stationery Office, www.legislation.hmso.gov.uk.

The Disability Discrimination Act 1995 (DDA) (1995) The Stationery Office, www.legislation.hmso.gov.uk.

The Football Stadia Improvement Fund and The Football Licensing Authority (2003) *Accessible stadia: A good practice guide to the design of facilities to meet the needs of disabled spectators and other users*. London.

The Stationery Office (TSO) *Accessible thresholds in new housing: guidance for housebuilders and designers*, The Stationary Office, 1999, ISBN 0-11-702333-7.

Water Supply (Water Fittings) Regulations 1999 (1999) *Guidance Document relating to Schedule 2: Requirements for water fittings*. SI 1999/1148.

Appendix 1

Comparisons of interpretations

Word or Phrase	ADM	BS8300
Access	Approach, entry or exit	Access to, and use of, facilities and egress except in cases of emergency
Accessible	With respect to buildings, such that people, regardless of disability, age or gender are able to gain access	Able to be accessed by disabled people
Accessible route		Any route that is used to approach, or move between or within a building, and is accessible to disabled people
Chair stairlift		Stairlift with a seat, which may be fixed or folding
Contrast visually	The difference in light reflectance between two surfaces is greater than 30 points. This is used to indicate the visual perception of one element of the building or a fitting within a building against another	
Dwelling	A house or a flat	
Effective clear width (BS8300) Clear opening width (ADM)	For dwellings, the clear opening width of a door is taken from the face of the door stop on the latch side to the face of the door when open at 90°	Available width measured at 90° to the plane of the doorway for passage through a door opening, clear of all obstructions, such as handles and weather boards on the face of a hinged door, when such a door is opened through 90° or more, or when a sliding or folding door is opened to its fullest extent

Word or Phrase	ADM	BS8300
Flight		Ramp or a continuous series of steps between two landings
Going		Horizontal distance between two consecutive nosings of a step, measured on the walk-line or the horizontal distance between the start and finish of a flight of a ramp
Handrail		Component of stairs, steps or ramps that provides guidance and support at hand level
Illuminance		Amount of light falling on a surface, measured in lumens per square metre (lm/m^2) or lux (lx)
Independent access	Access to a part of a building (from outside, and therefore from the site boundary and from any car park on site) that does not pass through the rest of the building	
Landing		Platform or part of a floor structure at the end of a flight or ramp, or to give access to a lift
Level	Being predominantly level but having a maximum gradient, along the direction of travel of 1 in 60. This applies to surfaces of a level approach, access routes and landings to steps, stairs and ramps	
Luminance		Brightness or light intensity of a surface, measured in candelas per square metre (cd/m^2) *Note*: Surfaces with different luminances can be distinguished from one another by people who are colour blind
Nosing		Projecting front edge of a tread or landing that may be rounded, chamfered or otherwise shaped

Word or Phrase	ADM	BS8300
Platform lift		Lift with a platform and low walls and which travels vertically between two levels and is intended for use standing up or seated on a chair or a wheelchair
Plot gradient	The gradient measured between the finished floor level of the dwelling and the point of access	
Point of access	(Dwellings) The point at which a person visiting a dwelling would normally alight from a vehicle, which may be inside or outside the boundary of the premises, prior to approaching the dwelling	
Principal entrance	The entrance which a visitor, not familiar with the dwelling, would normally expect to approach, or the common entrance to a block of flats	Entrance to a building which a visitor would normally expect to approach
Ramp		Construction, in the form of an inclined plane 1:20 or steeper from the horizontal or a series of such planes and an intermediate landing or intermediate landings that make it possible to pass from one level to another
Rise		Vertical distance between the upper horizontal surfaces of two consecutive treads, or of a landing and the next tread above or below it, or of a flight between two consecutive landings
Riser		Vertical component of a step between tread or landing or the tread or landing above or below it
Spillover		Interference within one induction loop from a signal from another induction loop nearby

Word or Phrase	ADM	BS8300
Stair clear width		Unobstructed minimum distance on plan perpendicular to the walking line of a stair
Stairlift		Lift that travels from one level to another along a line parallel with the pitch line of the stair
Steeply sloping plot	(Dwellings) A plot gradient of more than 1 in 15	
Tactile paving		Profiled paving surface providing guidance or warning to blind and partially sighted people
Tread		Horizontal component of a step
Unisex		Facility designed for use by either sex with or without assistance by people of the same or opposite sex
Wheelchair stairlift		Stairlift with a horizontal platform which accommodates a wheelchair user

Appendix 2

Example extracts from Access Audits

The following sections are examples taken from an access audit. The priority column gives the priorities in this report as:

Priority Key:

1 Access or use of service impossible or unreasonably difficult, therefore contrary to DDA (highest priority)

2 Access or use of service difficult, alter as soon as possible. Good practice.

3M Alter as part of normal maintenance works i.e. when maintenance is required then change for a better solution.

3R Alter as part of refurbishment i.e. when refurbishment is required then change for a better solution.

4 Alter or improve when a specific need is identified e.g. when an employee needs it changing.

5 No action reasonably practical.

By use of the Guidance Clause column, the service provider can understand where the auditor gathered the information for his comment, and can then review any changes in the light of the guidance clause given, or compare with other guidance.

The following pages show example extracts from Access Audits.

Item	Sub Item	Observation	Guidance Clause	Recommendation	Priority
3.3 Entering a building (cont)	3.3.4 Door size	The principal entrance doors are single swing, inward opening doors providing an effective clear width of 710 mm per leaf. This can be a hindrance to wheelchair users trying to negotiate the doors	BS8300 clause 6.4.1& 6.4.2	Consideration should be given to providing a principal entrance door having an effective clear width of 800 mm and an unobstructed leading edge of at least 300 mm	1
	3.3.5 Opening force	The principal entrance door has an opening resistance in excess of 20 N	BS8300 clause 6.3	Reduce opening resistance of door to 20 N or less	1
	3.3.6 Vision	Vision should be provided to main entrance doors to alert people approaching the door of the presence of someone on the other side There are vision panels provided to the principal entrance doors at a height between 1350 and 1850 mm. This is too high to be useful for wheelchair users or people of small stature. See photograph	BS8300 clause 6.4.3	Provide suitable vision at a height of between 500 and 1500 mm of the floor to external entrance door	1
	3.3.7 Door furniture	It should be possible to operate door handles with one hand and without having to tightly grasp the handle or twist the wrist The principal door is fitted with a silver lever handle at a height of 1050 mm above the floor	BS8300 clause 6.5.1	The silver-coloured handle currently contrasts well with the green door. If the door were to be repainted, as a solution to the issue discussed at 3.3.1, adequate contrast of the handle should be maintained	3R

Item	Sub Item	Observation	Guidance Clause	Recommendation	Priority
	3.3.8 Entrance lobby	The principal entrance doors open into a 3480 mm deep × 1750 mm wide entrance foyer	BS8300 clause 6.3.6	None	—
3.4 Horizontal circulation	3.4.1 Entrance hall and reception area	There is no formal reception area for this building. The entrance to the main Bar/Seating/Stage area is immediately apparent upon arrival	BS8300 clause 11	None	—
3.9 Individual rooms	3.9.1 Toilet accommodation	Disabled people should be able to access and use toilet accommodation	BS8300 clause 12.4	Accessible WC facilities should be provided	1
		There are separate male and female toilets provided within this building The toilets are not accessible or usable by wheelchair users and have no provision for ambulant disabled people		Consideration should be given to improving the existing toilet areas so as to make it accessible and usable by wheelchair users. Figure 55, BS8300 gives details of a suitable unisex accessible toilet layout (including door. size/configuration). sufficient It is suggested that there is space within the Gents WC area to form an accessible unisex WC, subject to suitable door widths etc from the foyer being provided	1
				If fully accessible unisex provision cannot be made then consideration should be given for provision for ambulant disabled people in the existing toilet – see BS8300 Figure 60	1

Item	Sub Item	Observation	Guidance Clause	Recommendation	Priority
	3.9.2 Kitchens/ refreshment areas	The kitchen area is fitted with standard worktops and sink unit at 900 mm above the floor. This could cause problems for any staff/volunteers that are wheelchair users	BS8300 clause 12.1	Kitchen and Refreshment areas would be more accessible if lowered areas were provided to allow wheelchair users to access refreshment facilities such as cooking and tea-making provision, etc.	4
	3.9.3 Beer cellar	The side entrance to the beer cellar is provided with a concrete ramp downinto the room from the external door. The ramp does not contrast well with the floor of the cellar and poses a trip hazard, especially for people with impaired vision	BS8300 clause 8.2.6	Consideration should be given to highlighting the presence of the ramp with suitable markings which contrast in luminance and colour with the surrounding floor/walls	1
	3.9.4 Stage area	The stage is set at a height of 350 mm above the adjacent dance and seating areas. The transition to the stage is provided with a visual contrast. However, the step could be a hindrance to wheelchair users and those with mobility problems participating in a performance on the stage	BS8300 clause 8.1.4	Consideration should be given to accessing the stage, for example: a. provide suitable ramp b. provide platform lift c. improving the handrail	4

Appendix 3

Workshops: some commonly queried scenarios

Workshop 1

Query regarding lifts into existing commercial buildings due to new access regulations

A property is being converted into a commercial building. Space is tight and the owner does not want to fit a lift.

Where work is being carried out on a property, this may invoke Building Regulations. In such a case, one would have to determine if Part M applies and to what extent. A discussion of the application continues below. If Part M applies, this is the primary consideration. Approved Document M (ADM) can be used for guidance and/or BS8300. If the latter, this needs to be clarified, discussed and argued in the Access Statement to be provided with the Building Regulation application. Following the completion of the works and occupation, Disability Descrimination Act (DDA) and the Fire Precautions (Workplace) Regulations take effect.

If no work, change of use, alterations or extensions are being carried out then the DDA remains the primary force and consideration will need to be given to the provision of service to the public, and employment issues.

Material change of use

Refer to the actual Building Regulations to determine the effect and requirements related to a material change of use. Regulation 5 (of Building Regulations 2000 as amended) is the relevant part. Follow the argument below:

A 'material change of use' includes:

> *(f) the building is not a building described in Class I to VI in Schedule 2 where previously it was (e.g. barn conversion)*
>
> *(j) the building is used as a shop where previously it was not*

Note that a change of use to an office is not a material change of use for Regulation 5.

Requirements for material change of use

Regulation 6: Parts of A, B, F, G, H, J, L

If (j) the above plus M1

So if a property was converted to a shop, then M1 (i.e. access and use of facilities) would kick in.

Extension or material alteration

Regulation 4: Where a building is extended or undergoes a material alteration, the work (and therefore the extension) must comply with M1 (i.e. access and use of facilities).

This means that alterations to features relevant to compliance with Part M1, such as entrances or arrangements for people to get from one level to another within the building must result in features that comply with M1.

So it can be argued, probably successfully, that if you put in new or extended floors, or bring into use previously unused floors, access must be achieved to those levels. Stairs are a means of access for some people, and lifts are another, therefore to comply with M1, both are needed, and likely to be necessary.

Do not forget means of escape for wheelchair users. Refuges are possibly not acceptable for the fire risk assessment as the fire authority state that they are not under a duty to necessarily rescue people from refuges. People should be able to escape under their own unimpeded efforts.

The point is clarified by ADM 0.14 (i) c:

> *"What requirements apply*
>
> *0.14 If Part M applies, reasonable provision should be made in:*
>
> *i) Buildings other than dwellings*
>
> ....

> *c. so that people, regardless of disability, age or gender, can have access into, and within, any storey of the building and to the building's facilities, subject to the usual gender-related conventions regarding sanitary accommodation;"*

In addition, an Access Statement should be prepared.

Workshop 2

Query regarding colour contrast and fonts

Signs need to be designed, for various reasons, to be positioned internally and externally to a building. What colour contrasts are not suitable?

Simple care when selecting fonts and colours can make the sign more accessible.

For example, there are different combinations, which can affect people with different forms of colour blindness or other visual disorders as Figure A3.1 shows.

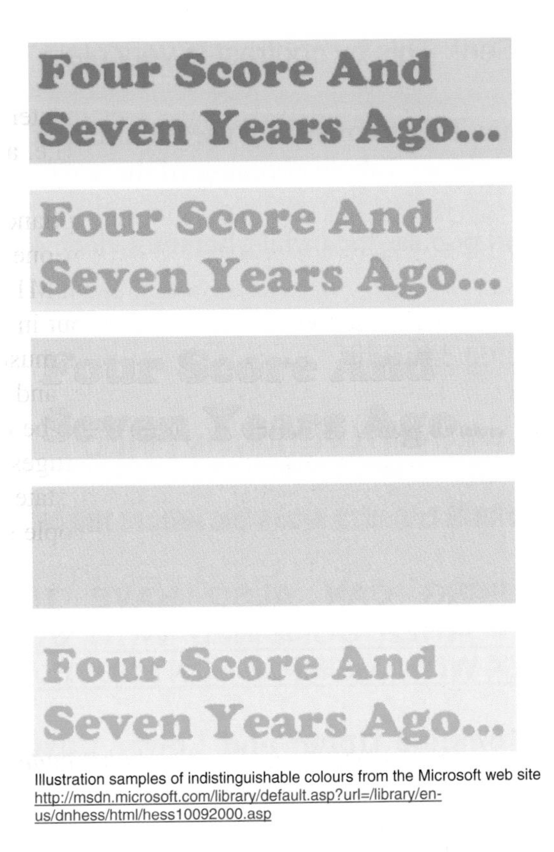

Figure A3.1 Colour contrasts for publications, signs, notices, presentations etc

Black, dark blue or dark green on white, or white on any of these colours is clear but there are many other suitable combinations.

Fonts

Fonts can also make a difference. The size of the font is important and depends on the purpose and distance from the reader. For written documentation, fonts less than 12 point should be avoided. Most people in their 40s begin to note an exponential decline in their reading range until their need to regularly wear reading glasses. Fonts of at least 12 point are helpful.

CommercialPrint BT This is 12 point but very difficult to read

Blackadder ITC: This is 12 point but extremely difficult to read.

Edwardian Script is very beautiful but impossible to read

Microsoft Sans Serif: This by contrast is very plain

Times New Roman: This a commonly used font but at 12 point is not as easy to read as Arial or Tahoma because of the serifs

Garamond: was also popular for a while, but the serifs make it less clear.

Tempus sans ITC is more modern, but still contains idiosyncrasies which makes reading difficult

AS DOES MATISSE ITC WHICH PUTS EVERYTHING IN CAPITALS

Impact: Too much bold can also make the letters run into each other.

AND UNDERLINING CAN ALSO HAVE THIS EFFECT PARTICULARLY WHEN COMBINED WITH BOLD AND/OR CAPITALS EVEN WITH A SIMPLE FONT SUCH AS ARIAL

Signs Which Combine Upper and Lower Case Letters Are Usually The Best

Figure A3.2 Fonts and styles

For most signs, unhelpful font types should be avoided. This would include italics and fonts with serifs. Serifs are the small embellishments to letters. Figure A3.2 shows various examples.

If you are wishing to impart information, make that information as easy to read as possible. Communication depends on the other person receiving the information in a form they can comprehend.

Examples of signboard colours

The colour and luminance of letters, symbols and pictograms should contrast with the colour and luminance of the signboard. Signboards themselves should contrast in colour and luminance with their backgrounds.

For example, boards positioned against bushes should not be dark green, boards on brick walls should not be dark red.

Light-coloured text and symbols or pictograms on a dark background are preferred.

The illustrations in Figure A3.3 are taken from Table 6 in BS 8300.

Further information about fonts for documents to be issued to the public are included in the RNIB document, "See it right – Getting your message across" and includes the advice given in Table A3.1.

Workshop 3

A redundant 19th century industrial building could be changed into flats. What are the implications of this scenario for Regulation M?

1. Examine the meaning of change of use in relation to Part M and the proposed scenario:

   - Regulation 5 of the Building Regulations 2000 relates to whether or not a change of use is material under the regulations

   - These are described (a) to (j) as below:
     (a) the building is used as a dwelling, where previously it was not;
     (b) the building contains a flat, where previously it did not;
     (c) the building is used as a hotel or boarding house, where previously it was not;
     (d) the building is used as an institution, where previously it was not;
     (e) the building is used as a public building, where previously it was not;
     (f) the building is not a building described in Classes I–VI in Schedule 2, where previously it was;
     (g) the building, which contains at least one dwelling, contains a greater or lesser number of dwellings than it did previously;
     (h) the building contains a room for residential purposes, where previously it did not;

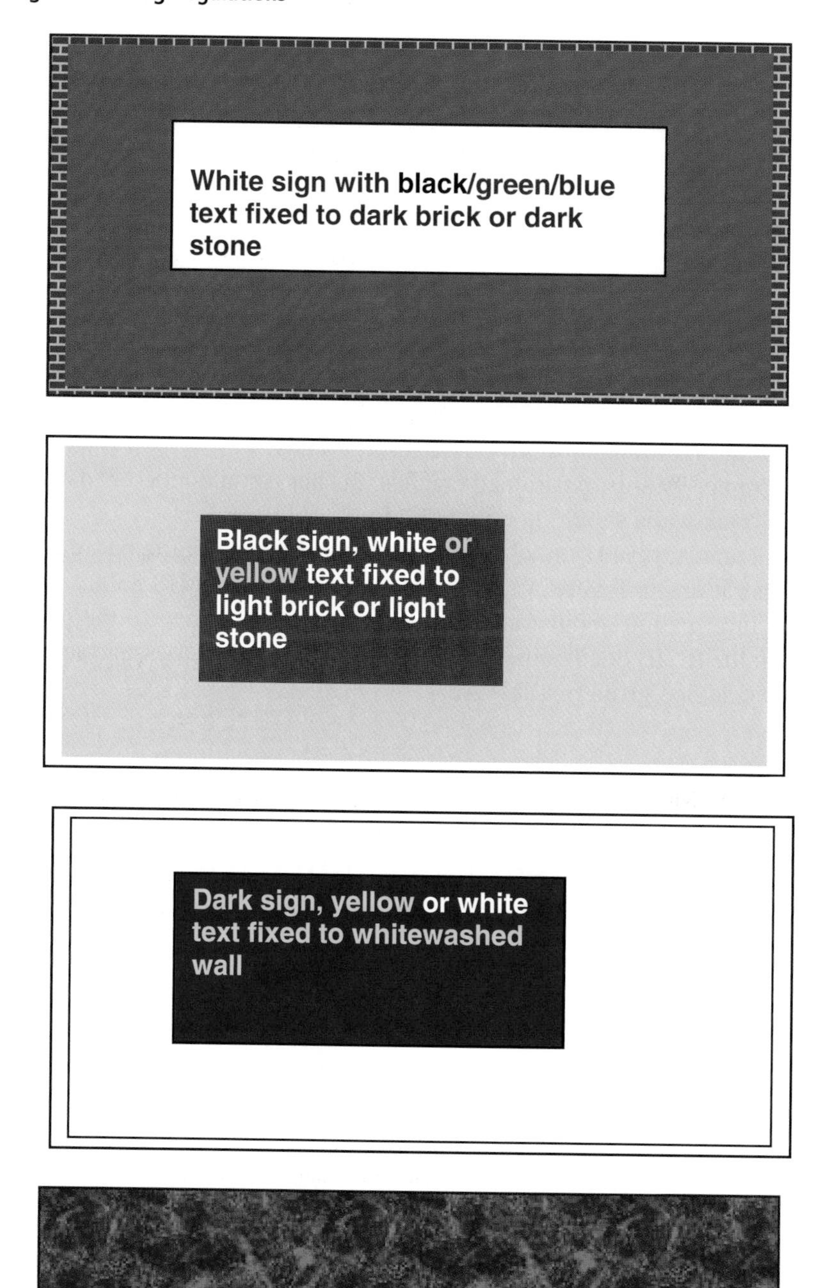

Figure A3.3 Sign board colours

Table A3.1 Type advice

Feature	Comment
Type size	12 point minimum, preferably 14 or 15 point
Typeface	Avoid italics and ornate typefaces
Type style	Avoid capitals for continuous text Lower case lettering with some initial capitals is generally easier to read Avoid underlining
Type weight	Avoid lightweight types, bold or semi-bold preferred
Leading (or spaces between lines of type)	Preferably 1.5–2 times the space between words on a line
Contrast	The better the contrast between the background and the type, the more legible it is
Numbers	If using numbers, choose a typeface in which the numbers are clear, particularly 3, 5, 8 and 0
Line length	Ideally the line length should be between 60 and 70 letters per line
Word spacing and alignment	Keep to the same amount of space between each word, it is best to avoid justified text
Columns	Make sure the space between columns clearly separates them; if space is limited use a vertical line
Reversing type	If using white type, make sure the background colour is dark enough to provide sufficient contrast
Setting type 1	Avoid fitting text around images if this means that lines of text start in a different place, and, therefore, are difficult to find
Setting type 2	Avoid vertically set text; avoid setting text over images, for example, photographs
Paper	Avoid glossy paper because glare makes it difficult to read; choose uncoated paper that weighs over 90 gsm

(i) the building, which contains at least one room for residential purposes, contains a greater or lesser number of such rooms than it did previously; or
(j) the building is used as a shop, where previously it was not.

- Regulation 6 states the requirements relating to each material change of use. These are summarised in Table A3.2.

Table A3.2 Requirements for material change of use

	Scenario, change of use to or to have:	The following parts of Building Regulations apply
(a)	A dwelling	B1, B2, B3, B4(2), B5, C4, E1, E2, E3, F1, F2, G1, G2, H1, H6, J1, J2, J3, L1, L2
(b)	A flat	B1, B2, B3, B4(2), B5, E1, E2, E3, F1, F2, G1, G2, H1, H6, J1, J2, J3, L1, L2
(c)	A hotel or boarding house	A1, A2, A3, B1, B2, B3, B4(2), B5, E1, E2, E3, F1, F2, G1, G2, H1, H6, J1, J2, J3, L1, L2, M1
(d)	An institution	A1, A2, A3, B1, B2, B3, B4(2), B5, F1, F2, G1, G2, H1, H6, J1, J2, J3, L1, L2, M1
(e)	A public building	A1, A2, A3, B1, B2, B3, B4(2), B5, E4 (for schools only), F1, F2, G1, G2, H1, H6, J1, J2, J3, L1, L2, M1
(f)	A building which was in Classes I – VI in Schedule 2, i.e., exempt from Building Regulations, which is no longer e.g., a converted farm building	A1, A2, A3, B1, B2, B3, B4(2), B5, F1, F2, G1, G2, H1, H6, J1, J2, J3, L1, L2
(g)	A building with an existing dwelling now with more/less dwellings	B1, B2, B3, B4(2), B5, E1, E2, E3, F1, F2, G1, G2, H1, H6, J1, J2, J3, L1, L2
(h)	Contains a residential room	B1, B2, B3, B4(2), B5, E1, E2, E3, F1, F2, G1, G2, H1, H6, J1, J2, J3, L1, L2
(i)	An existing residential with more/less residential rooms	B1, B2, B3, B4(2), B5, E1, E2, E3, F1, F2, G1, G2, H1, H6, J1, J2, J3, L1, L2
(j)	A shop	B1, B2, B3, B4(2), B5, F1, F2, G1, G2, H1, H6, J1, J2, J3, L1, L2, M1

- A change of use from an industrial building to flats comes under subsection (b). The requirements relating to the change of use do not include Part M. Therefore, there is no requirement to apply Part M to the conversion.

2. Consider whether the associated works (e.g. new corridors, new dwelling entrances and new sanitary accommodation) need to comply with part M.

   - Regulation 3 of the Building Regulations 2000 states that an alteration is material if it results in a building which does not comply when previously it did, or, if it did not comply before, it is more unsatisfactory after.

This is the debatable part. Take, for example, the provision of a dwelling in the building. The dwelling does not comply with Part M as guided by ADM 2004 in that the entrance door is not accessible (Paragraph 6.23), the corridors are too narrow (Section 7) and the WC is not accessible (Section 10).

- The standard of compliance is now more unsatisfactory because people are more likely to wish to access these facilities and may be unable to do so.
  or
- There were no facilities before, so any standard is better than none.

The pragmatic approach may be to ensure any new work complies with recommendations in ADM as the costs will probably add little compared to the total cost of the new works.

3. Consider whether existing, and unchanged access measures need to be altered or enhanced. For example, should a lift be provided? Should the stairs be upgraded to ambulant accessible standard?

- With reference to the change of use considerations of Regulation 5 and the material alteration considerations in Regulation 3, if the stairway in the building is existing and unaltered, it would seem that there is no requirement to alter it under Building Regulations.

- The Approved Document in Section 9 does not put the provision of a lift, even in newly built flats, as being essential. The provision of a lift in a refurbished building would therefore not be a requirement of Building Regulations.

Index

Note: Italicized page numbers denote corresponding entries are taken from tables.